T0315373

# Advances in Welding Technologies for Process Development

# Advances in Welding Technologies for Process Development

Edited by
Jaykumar J. Vora and Vishvesh J. Badheka

CRC Press
Taylor & Francis Group
Boca Raton London New York

CRC Press is an imprint of the
Taylor & Francis Group, an **informa** business

MATLAB® is a trademark of The MathWorks, Inc. and is used with permission. The MathWorks does not warrant the accuracy of the text or exercises in this book. This book's use or discussion of MATLAB® software or related products does not constitute endorsement or sponsorship by The MathWorks of a particular pedagogical approach or particular use of the MATLAB® software.

CRC Press
Taylor & Francis Group
6000 Broken Sound Parkway NW, Suite 300
Boca Raton, FL 33487-2742

First issued in paperback 2020

© 2019 by Taylor & Francis Group, LLC
CRC Press is an imprint of Taylor & Francis Group, an Informa business

No claim to original U.S. Government works

ISBN 13: 978-0-367-65651-5 (pbk)
ISBN 13: 978-0-8153-7707-8 (hbk)

This book contains information obtained from authentic and highly regarded sources. Reasonable efforts have been made to publish reliable data and information, but the author and publisher cannot assume responsibility for the validity of all materials or the consequences of their use. The authors and publishers have attempted to trace the copyright holders of all material reproduced in this publication and apologize to copyright holders if permission to publish in this form has not been obtained. If any copyright material has not been acknowledged please write and let us know so we may rectify in any future reprint.

Except as permitted under U.S. Copyright Law, no part of this book may be reprinted, reproduced, transmitted, or utilized in any form by any electronic, mechanical, or other means, now known or hereafter invented, including photocopying, microfilming, and recording, or in any information storage or retrieval system, without written permission from the publishers.

For permission to photocopy or use material electronically from this work, please access www. copyright.com (http://www.copyright.com/) or contact the Copyright Clearance Center, Inc. (CCC), 222 Rosewood Drive, Danvers, MA 01923, 978-750-8400. CCC is a not-for-profit organization that provides licenses and registration for a variety of users. For organizations that have been granted a photocopy license by the CCC, a separate system of payment has been arranged.

**Trademark Notice:** Product or corporate names may be trademarks or registered trademarks, and are used only for identification and explanation without intent to infringe.

---

**Library of Congress Cataloging-in-Publication Data**

---

Names: Badheka, Vishvesh J., editor. | Vora, Jaykumar, editor.
Title: Advances in welding technologies for process development / Vishvesh J. Badheka and Jaykumar Vora, editors.
Description: First edition. | Boca Raton, FL : CRC Press/Taylor & Francis Group, 2019. | Includes bibliographical references and index.
Identifiers: LCCN 2018047805| ISBN 9780815377078 (hardback : acid-free paper) | ISBN 9781351234825 (ebook)
Subjects: LCSH: Welding.
Classification: LCC TS227 .A3236 2019 | DDC 671.5/2—dc23
LC record available at https://lccn.loc.gov/2018047805

---

**Visit the Taylor & Francis Web site at**
**http://www.taylorandfrancis.com**

**and the CRC Press Web site at**
**http://www.crcpress.com**

# Contents

Preface...........................................................................................................vii
Foreword .........................................................................................................xi
Acknowledgments .........................................................................................xiii
Editors.............................................................................................................xv
Contributors.................................................................................................. xvii

1. Multiple-Wire Welding GMAW and SAW.............................................1
   *Syed Quadir Moinuddin and Abhay Sharma*

2. Regulated Metal Deposition (RMD™) Technique for Welding
   Applications: An Advanced Gas Metal Arc Welding Process............23
   *Subhash Das, Jaykumar J. Vora, and Vivek Patel*

3. Friction Stir Welding and Its Variants .......................................................33
   *Vivek Patel, Wenya Li, and Devang Sejani*

4. Different Methodologies for the Parametric Optimization of
   Welding Processes .......................................................................................55
   *Jaykumar J. Vora, Kumar Abhishek, and PL. Ramkumar*

5. Additive Manufacturing with Welding....................................................77
   *Manish Kumar, Abhay Sharma, Uttam Kumar Mohanty, and
   S. Surya Kumar*

6. Cares to Deal with Heat Input in Arc Welding: Applications
   and Modeling ............................................................................................. 101
   *Américo Scotti*

7. Friction Stir Welding of High-Strength Steels.................................... 139
   *R. Ramesh, I. Dinaharan, and E. T. Akinlabi*

8. Friction Stir Spot Welding of Similar and Dissimilar
   Nonferrous Alloys..................................................................................... 159
   *R. Palanivel, I. Dinaharan, and R. F. Laubscher*

9. Linear Friction Welding ......................................................................... 191
   *Xinyu Wang, Wenya Li, Tiejun Ma, and Achilles Vairis*

10. Joining of Dissimilar Materials Using Friction Welding................... 211
    *Udaya Bhat K. and Suma Bhat*

**11. Insights into the Flux-Assisted TIG Welding Processes** ..................... 237
*Jaykumar J. Vora*

**12. Friction Stir Welding of Aluminum Alloy: Principle,
Processing, and Safety** ............................................................. 261
*S. T. Azeez and E. T. Akinlabi*

**Index** ....................................................................................... 281

# *Preface*

While manufacturing is taking a paradigm shift, need for cost-effective, eco-friendly, and productive welding processes have arisen. To suffice the parameters of cost, quality, timeliness, and ease of using, a number of innovations and developments have taken place in welding processes within the past two decades. The advancements carried out enabled the industry to easily weld the materials which were previously thought of difficult to weld or even unweldable. Many of the developments also enhanced the productivity of the existing conventional welding processes by combining multiple welding processes or accurate process parameter settings. Each chapter, in this volume, attempts to provide a qualitative description of the technology and practice specific to the manufacturing technique or techniques used and the viable outcomes.

This book gives an overview of recently developed welding processes as well as small or significant efforts done for the advancement in making the welding operation achievable with required optimum quality. The book comprises of advancements in both arc-assisted and solid-state welding processes. The book has been so comprised that each chapter has been written and the data have been presented in the simplest form to enable readers of all backgrounds, a clear and concise presentation. The issues discussed also have been written convincingly and clearly.

This book has a total of 12 chapters. Chapter 1 deals with the advancements in the most widely used gas metal arc welding (GMAW) process as well as submerged arc welding (SAW) processes, in which more than one wire were used to increase the productivity by increasing the deposition rate. The technique is particularly useful for the thigh thickness joints widely used in pressure vessel industry.

Chapter 2 deals with a fairly novel but less-experimented regulated metal deposition (RMD) technique, which is basically a controlled short-circuiting GMAW process. The technique is a potential candidate for the root pass joints of the high-thickness plates for achieving better results.

Chapter 3 deals with the various advancements taken place in the friction stir welding (FSW) process in terms of defining its new variants for several targeted applications.

Chapter 4 covers lucid and informative details about the various approaches experimented by the researchers to set the accurate process parameters for the welding processes and achieve targeted weld bead profile and subsequent properties.

Chapter 5 deals with a very novel topic of additive manufacturing with welding, in which recent advancements of the technique have been discussed.

Chapter 6 deals with a very important procedure which is common to all the welding process, i.e., heat input. The different methods of measuring and understanding the heat input have been discussed in more details.

Chapter 7 deals with the some advancements taken place in the FSW process which is particularly carried out to join high-temperature strength materials. Various adjustments such as tool material and fixtures have been discussed in order to enable the process to join these steels in the solid state.

Chapter 8 covers a widely used variant of FSW process termed as friction stir spot welding (FSSW), which is particularly developed for the replacement of resistance spot welding (RSW) carried out in lap configuration for low thickness.

Chapter 9 discusses about the linear friction welding process—a solid-state welding process—and highlights its performance vis-à-vis other welding processes.

Chapter 10 covers a detailed survey to address an imperative challenge in the industry pertaining to the joining of dissimilar materials having a varied melting point as well as widely different properties with FSW.

Chapter 11 deals with some of the advancements taken place in the tungsten inert gas welding (TIG) process by adding a flux. The chapter summarizes the different techniques and also compares the performance as well as the applicability of the same.

Chapter 12 deals with an important aspect in the field of safety of the FSW process, which is anyhow applicable to almost industries using the process.

Overall, we firmly believe that the book will be a useful addition to the knowledge database for the advanced welding processes in the domain. The book would be able to make the reader understand the past work, and idea seed for the process. The need for the advancement and the present status of the process along with concluding the applicability of the same has been thoroughly presented.

The book can be a one-point literature which can be referred by the students, researchers as well as industrial fellows working in the field to get an idea about the recent developments taken place in the past few decades.

We hope the contribution would be received wholeheartedly by the welding community positively.

**Jaykumar J. Vora**
**Vishvesh J. Badheka**

MATLAB® is a registered trademark of The MathWorks, Inc. For product information, please contact:

The MathWorks, Inc.
3 Apple Hill Drive
Natick, MA 01760-2098 USA
Tel: 508-647-7000
Fax: 508-647-7001
E-mail: info@mathworks.com
Web: www.mathworks.com

# *Foreword*

Manufacturing is one of the key parameters that plays a pivotal role in the development of a nation. Constant evolution and advancement in the different manufacturing processes have always been at the center of any activity due to the high expectations from these manufacturing processes in terms of yielding trustworthy components cost-effectively. Welding is one of the most widely used fabrication methods particularly for the production of automobiles, vessels for petroleum refining, boilers, heat exchangers, etc. This is the reason for the intensification of research and development activities pertaining to welding and joining of different similar and dissimilar combinations of the metals in the past decade. There are certain advances in the field of welding which have yielded excellent results and hence enhanced the productivity of the existing process. *Advances in Welding Technologies for Process Developments* primarily addresses the recent advances which have taken place in various welding processes across the domain to update the reader about the working principle, predicaments in the existing process, innovations did to overcome the limitations, and its direct industrial/practical application.

**Jaykumar J. Vora**
**Vishvesh J. Badheka**

# Acknowledgments

At the start, the editors would like to acknowledge the publisher CRC Press for believing in our idea and giving us a chance to work on bringing out a book on the topic. We also acknowledge the inputs and suggestions of the reviewers of the initial proposal for showing the pathway and refining our idea which further helped to cover the most relevant topics of the book.

The editors appreciatively acknowledge the considerate and wholehearted support received from the authors of the chapters in this book and especially their intent to disseminate their knowledge with the world through the chapter. The efforts made by the contributing authors to effectively analyze and present various topics in their chapter for enhancing the quality as well as scientific know-how and technological advancement of the fraternity at large is greatly acknowledged. The reviewers of the chapters are also acknowledged for devoting their valuable time and energy in giving their views and insights for enhancing the quality of the chapter. Their selfless support has really been a pathway for improvement.

We also extend an wholehearted thanks and the deepest gratitude to the editorial staff of CRC Press and especially Dr. Gagandeep Singh (Editorial Manager, Senior Editor Engineering/Environmental Science) and Ms. Mouli Sharma (Editorial Assistant) for their constant motivation, interest, involvement, attention, and their willingness to assist even at short notice for the questions from the editors as well as the contributing authors. Their support in testing times, understanding the constraints, and views on resolving the same was indeed impeccable. Thus, in a nutshell, the book is at this stage only due to three strong support pillars, namely, contributing authors, reviewers, and the editorial staff.

Last but not the least, the support and guidance from family, seniors, and colleagues at the times in need are also acknowledged. Family and friends are always a source of inspiration and support who are also acknowledged.

Jaykumar J. Vora
Vishvesh J. Badheka

# Editors

**Jaykumar J. Vora** is an assistant professor in the Mechanical Engineering Department of Pandit Deendayal Petroleum University (PDPU), Gandhinagar, Gujarat, India. He pursued his master's in Welding Technology from The Maharaja Sayajirao University of Baroda (MSU) as a sponsored candidate from M/s Larsen and Toubro Ltd. Subsequently, he completed his Ph.D. in the area of advanced welding technology from PDPU. He is recognized as a *Chartered Engineer—India* and is the recipient of the prestigious Visiting Scientist Award from Indian National Science Academy (INSA). He has also worked as a visiting faculty and research scholar at Lamar University, Texas, USA. He is empaneled as a subject matter expert on various industrial organizations such as Tata Consultancy Ltd. and Centre for Entrepreneurship Development (Govt. of Gujarat). Prior to joining PDPU, he worked with industries such as L&T, and Tata Motors. Currently, he is guiding three Ph.D. students from industries along with master's and bachelor's students.

He has won several best paper awards in various conferences, and his research team was awarded second place in IIW young engineers' competition in 2017. Actively involved in the research in close correlation with industries and labs, he is pursuing the development of advanced welding techniques such as A-TIG, FSW, FSSW, and LBW, particularly for low alloy steels, dissimilar combinations as well as different ferrous and nonferrous steels. Owing to the INSA award, he worked at ARCI, Hyderabad for a brief period in the field of development of A-LBW and LBW process. He has several international journal publications and attended several national and international conferences with the support of SERB, Govt. of India. He is also the reviewer for many international peer-reviewed journals/books and chaired several technical sessions in international conferences. Currently, he is working on techniques to incorporate nanotechnology in welding processes for improving their performance for which he is in the process to file a patent. In addition to this, recently he has created an online YouTube channel named "Researchers Online" wherein he has uploaded several videos covering basic aspects of research, particularly for young researchers and students.

 **Vishvesh J. Badheka** studied metallurgical engineering at The M.S. University of Baroda. He received bachelor's, master's, and Ph.D. in Metallurgical Engineering. He qualified Diploma International Welding Engineer awarded by International Institute of Welding in December 2011.

Currently, he is working as professor and head of Mechanical Engineering at School of Technology, Pandit Deendayal Petroleum University (PDPU), Gandhinagar, India. He received funding from various funding agencies including DST, ISRO, DRDO, and DAE for his research in the area of advanced welding processes includes, Narrow Gap Welding using metal-cored and flux-cored wires, friction stir welding similar and dissimilar metal, friction stir processing for surface composites and super plasticity, friction welding dissimilar metal combinations, and ATIG of reduced activation martensitic steels & 9Cr-1Mo steels. So far, he has completed five sponsored R&D projects and four ongoing projects. So far, he has guided 27 M.Tech. and 7 Ph.D. researches into his credit in the area of advanced welding processes. He has 40 international journal publications and 60 international and national conferences. He is also the member of various professional bodies such as IIW, ISNT, ISTE, IIM, IIF, MRSI, and AWS.

He has received numerous awards, including Dr Vikram Sarabhai Award for the year 2009–2010 for "Friction Stir Welding of Aluminum and its alloys"; from Gujarat Council on Science and Technology (GUJCOST), Government of Gujarat. He was also awarded a bronze medal by the "Center for Innovation, Incubation and Entrepreneurship (CIIE) of Indian Institute of Management, Ahmedabad (IIM)". He received several best oral presentation and best research paper awards by The Indian Institute of Welding.

# Contributors

**Kumar Abhishek**
Department of Mechanical
Engineering
Institute of Infrastructure
Technology Research and
Management
Ahmedabad, Gujarat, India

**E. T. Akinlabi**
Department of Mechanical
Engineering Science
University of Johannesburg
Johannesburg, South Africa

**S. T. Azeez**
Mechanical Engineering Science,
Faculty of Engineering and the
Built Environment
University of Johannesburg
Johannesburg, South Africa

**Suma Bhat**
Department of Mechanical
Engineering
Srinivasa School of Engineering
Srinivasanagar, Karnataka, India

**Subhash Das**
Welding Group
ITW India Pvt. Limited
Vadodara, Gujarat, India

and

Mechanical Engineering
Department
Pandit Deendayal Petroleum
University
Gandhinagar, Gujarat, India

**I. Dinaharan**
Department of Mechanical
Engineering Science
University of Johannesburg
Johannesburg, South Africa

**Manish Kumar**
Department of Mechanical and
Aerospace Engineering
Indian Institute of Technology
Hyderabad
Hyderabad, Telangana, India

**S. Surya Kumar**
Department of Mechanical and
Aerospace Engineering
Indian Institute of Technology
Hyderabad
Hyderabad, Telangana, India

**R. F. Laubscher**
Department of Mechanical
Engineering Science
University of Johannesburg
Johannesburg, South Africa

**Wenya Li**
State Key Laboratory of
Solidification Processing,
Shaanxi Key Laboratory
of Friction Welding
Technologies
Northwestern Polytechnical
University
Xi'an, PR China

**Tiejun Ma**
State Key Laboratory of
   Solidification Processing,
   Shaanxi Key Laboratory
   of Friction Welding Technologies
Northwestern Polytechnical
   University
Xi'an, PR China

**Uttam Kumar Mohanty**
Department of Mechanical and
   Aerospace Engineering
Indian Institute of Technology
   Hyderabad
Hyderabad, Telangana, India

**Syed Quadir Moinuddin**
Department of Mechanical and
   Aerospace Engineering
Indian Institute of Technology
   Hyderabad
Hyderabad, Telangana, India

**R. Palanivel**
Department of Mechanical
   Engineering Science
University of Johannesburg
Johannesburg, South Africa

**Vivek Patel**
Mechanical Engineering Department
Pandit Deendayal Petroleum
   University
Gandhinagar, Gujarat, India

and

State Key Laboratory of
   Solidification Processing, Shaanxi
   Key Laboratory of Friction
   Welding Technologies
Northwestern Polytechnical
   University
Xi'an, PR China

**R. Ramesh**
Department of Mechanical
   Engineering
PSG College of Technology
Coimbatore, Tamil Nadu, India

**PL. Ramkumar**
Department of Mechanical
   Engineering
Institute of Infrastructure
   Technology Research and
   Management
Ahmedabad, Gujarat, India

**Américo Scotti**
Division of Welding Technology
University West
Trollhättan, Sweden

and

Center for Research and
   Development of Welding
   Processes
Federal University of Uberlandia
Uberlandia, Brazil

**Devang Sejani**
Material Science
Technische Universität Darmstadt
Darmstadt, Germany

**Abhay Sharma**
Department of Mechanical and
   Aerospace Engineering
Indian Institute of Technology
Hyderabad, Telangana, India

**Udaya Bhat K.**
Department of Metallurgical and
   Materials Engineering
National Institute of Technology
   Karnataka
Srinivasanaga, Karnataka, India

**Achilles Vairis**
State Key Laboratory of
   Solidification Processing, Shaanxi
   Key Laboratory of Friction
   Welding Technologies
Northwestern Polytechnical
   University
Xi'an, PR China

and

Mechanical Engineering Department
TEI of Crete
Heraklion, Greece

**Jaykumar J. Vora**
Mechanical Engineering
   Department
Pandit Deendayal Petroleum
   University
Gandhinagar, Gujarat, India

**Xinyu Wang**
State Key Laboratory of
   Solidification Processing,
   Shaanxi Key Laboratory of
   Friction Welding Technologies
Northwestern Polytechnical
   University
Xi'an, PR China

# 1

## Multiple-Wire Welding GMAW and SAW

**Syed Quadir Moinuddin and Abhay Sharma**

*Indian Institute of Technology Hyderabad*

### CONTENTS

1.1  Introduction ........................................................................................ 1
1.2  Background ......................................................................................... 2
1.3  Multiple-Wire Technology ............................................................... 3
    1.3.1  Twin-Wire Welding ............................................................... 4
    1.3.2  Welding with Three or More Wires ..................................... 7
1.4  Recent Advancements in Multiple-Wire Technology ................... 8
    1.4.1  Tandem GMAW ...................................................................... 8
    1.4.2  Power Source Synchronization .......................................... 10
    1.4.3  TIME-Twin ............................................................................ 11
    1.4.4  Double-Electrode GMAW and SAW ................................. 12
    1.4.5  Twin-Wire Indirect Arc Welding ...................................... 13
    1.4.6  Cable-Type Welding Wire Arc Welding ........................... 14
    1.4.7  Metal Powered-Assisted Multiple-Wire Welding ........... 15
1.5  Recent Trends in Research in Multiple-Wire Welding ............... 16
1.6  Conclusions ...................................................................................... 17
References ................................................................................................... 18

## 1.1 Introduction

Welding is a well-established joining technique used in industries for various applications. This technique is classified based on various factors such as use or no use of filler wire, fusion or solid-state welding, and shielding medium (gas or flux) (Eisenhauer 1998). The fusion welding is the most frequently used process wherein an electric arc is generated between a workpiece and an electrode. The resulting arc heat is used to melt the electrode wire and the material. The submerged arc-welding (SAW) and gas metal arc-welding (GMAW) processes are widely used in industries because of higher deposition rates and economic efficiency. In the past few decades, a significant advancement in both the processes has taken place.

The requirement of deposition and production rates at the shop floor is quite higher than what is possible with single-wire welding. In order to overcome these limitations, research is being conducted on existing technologies and materials as well as on new materials and technologies. The multiple-wire welding is one of such techniques that came into existence to increase the productivity. This process finds application where a high deposition rate without deep penetration is beneficial. Heavy corner welds, surfacing deposits, and different butt and fillet welds, where fit-up or melt-thru is critical, are some of the examples. Initial applications were focused on twin-wire welding and later with the help of advancements in torch configuration (triple-wire, four-wire, and six-wire) and power source, where several wires were connected to single or individual power sources. Various techniques have been developed for faster deposition: twin-wire welding, tandem welding, transferred ionized molten energy (TIME), double-electrode indirect arc welding, cable-type welding, etc. The difference between twin-wire and tandem welding is based on contact tube arrangement, i.e., common contact tube is used in twin-wire, whereas separate contact tubes are used in single or different torch/s in tandem welding. The double-electrode process consists of two torches: a main torch and a bypass torch. The bypass torch is used to provide additional melting without affecting the base metal, thus increasing the deposition rate. Indirect arc-welding process redistributes the energy between the wires, and the base plate is independent of the current circuit. The arc burns between the wires and increases the deposition rate. In the cable-type welding process, the wires are twisted around the center. In this process, arc rotation phenomenon causes an increase in deposition rate and reduces the arc pressure forming shallow molten pool. A considerable progress—academic and industrial—in the past decade has taken place in multiple-wire welding, which needs consolidation for the use of practitioners and researchers. The current chapter presents the recent developments in multiple-wire welding from the technology and research perspectives.

## 1.2 Background

The idea of multiple-wire welding in the form of twin-wire welding was first proposed by Ashton (1954). The references of multi-wire welding can be found in the earlier works of the 1950s and 1960s (Mandelberg and Lopata 1966; Uttrachi 1966). The interest in this technology started to grow in the 1980s with the initiation of the shop-floor application of the process. Thereafter, several developments have been reported (Chandel 1990; Tusek and Suban 2003; Sharma et al. 2008a,b, 2009a–c, 2015; Shi et al. 2014a; Yu et al. 2016). The initial research studies were mainly related to process establishment and feasibility. With the development of advanced power sources and modeling and simulation techniques, the research interest has grown in topics like melting rate (Tusek 2000), heat transfer, fluid flow modeling (Sharma et al. 2005a,b), etc. In the late 20th century, the idea of the multiple-wire was first applied in the GMAW process by Lassaline et al. (1989). Thereafter, twin and tandem wire with the GMAW was also reported (Suga et al. 1997; Michie et al. 1999). In the past decade, research interest in multiple-wire GMAW has grown due to the availability of advanced power sources that provide precise control and modulation of the welding arc (Scotti et al. 2006; Chen et al. 2016; Moinuddin et al. 2016; Ueyama et al. 2005b; Reis et al. 2015). The multiple-wire welding is expected to make a significant impact in the coming time that necessitates a consolidation of existing technologies and future possibilities. In order to understand the process, the following section briefs about the multiple-wire welding technique followed by the recent developments.

## 1.3 Multiple-Wire Technology

The multiple-wire welding uses more than one electrode (wires in the case of consumable arc welding), in which the electrodes travel simultaneously with the same or different feed rates through a joint contact tube. The wires may have a joint or individual feed rate control based on the application of one or more power sources. The wires in the contact nozzle can be arranged in parallel, series, or to form an optional shape (e.g., triangle) with regard to the welding direction (Tusek and Suban 2003). The multiple-wire operations cover the following three techniques:

- *Multiple Electrodes, Parallel Power*: In this system, multiple electrodes are connected with the same power source.
- *Two Electrodes, Series Power*: In this system, each electrode operates independently and remains isolated to each other. Each electrode possesses a feed motor and a voltage control. The power supply cable is connected to one welding head, and the return cable is connected to the second welding head instead of the workpiece.
- *Multiple Electrodes, Multiple Powers*: In this system, two or more electrodes are connected to individual power sources. The electrodes are separately connected to AC or DC power sources.

A classification, specifically for multi-wire GMAW/SAW, based on power source arrangement and contact tip configuration (Blackman et al. 2002) is suggested as given in Table 1.1.

**TABLE 1.1**

Classification of Multi-Wire for GMAW/SAW Systems

| Description Power | Wire Feeding Unit | Contact Tip Configuration | Transfer Modes |
|---|---|---|---|
| Twin-wire with single power source | A single-wire feed unit modified to feed two wires | A single common potential contact tip | Pulsed |
| Twin-wire with two power sources | Two-wire feed units that feed two wires | A single common potential contact tip | Pulsed |
| Asynchronous tandem | Two-wire feed units that are not directly synchronized | Two electrically isolated contact tips | Dip, pulsed, spray |
| Synchronized tandem | Two-wire feed units with synchronized directly | Two electrically isolated contact tips | Dip, pulsed, spray |

In view of the significant developments in the past few years, a comprehensive classification of multiple-wire welding based on the torch, contact tube, power sources, polarity, and welding directions is shown in Figure 1.1.

In multiple-wire welding, the wire arrangement within the contact tube and welding direction play important roles. In the case of wire in series, the first wire melts the parent metal. The weld pool thus produced stays behind the first arc due to high welding velocity and rearward motion of weld metal. The remaining wire/s, traveling behind the first one, shape/s the weld. The first wire, due to lower resistance, conducts a higher current than the other wires (Tusek and Suban 2003). Apart from this, the process parameters—welding current, voltage, wire diameter, the distance between two wires, contact tube-to-work piece distance (CTWD), welding speed, shielding gas, flux height, etc.—play a vital role.

### 1.3.1 Twin-Wire Welding

The twin-wire welding is the most investigated and frequently used variant of the multiple-wire technology. The term "twin-wire", or sometimes "twin-electrode", is characterized by a twin bore contact tube (as shown in Figure 1.2), in which two wires are fed through a single welding torch into a common weld pool (Tusek 2000).

In some of the earlier references like Mandelberg and Lopata (1966), a two-wire and two-power source type of system is described as the twin-wire welding process. In due course of time, a clear definition emerged for twin-wire welding. As categorically mentioned by Hinkel and Frosthoefel (1976), "Twin-electrode SAW, for those who are not familiar with the system, differs significantly from tandem arc welding, but only slightly from the single arc operation." Instead of feeding one electrode, the feeder is equipped with a double set of drive rolls and guides so that two electrodes can be fed simultaneously at the same feed rate. Both of these electrodes are normally

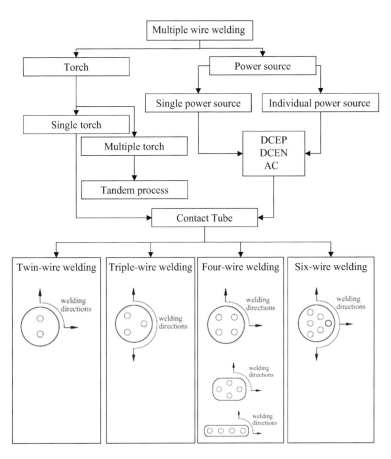

**FIGURE 1.1**
Classification of multiple-wire welding (Tusek and Suban 2003). (With kind permission from Elsevier.)

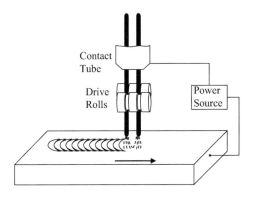

**FIGURE 1.2**
Schematic diagram of twin-wire welding (Sharma et al. 2008a,b). (With kind permission from Springer.)

fed through a common contact tip or set of jaws, so that the current from the power source splits between the two electrodes. Thus, a clear definition for twin-electrode welding, now more-often termed as "twin-wire welding", clearly exists. Furthermore, the process with two electrodes with individual power sources, where one electrode is following the other, is termed as tandem welding.

In twin-wire welding, the two wires are positioned in-line or at any angle across the seam with a single or multiple power sources. Figures 1.3 and 1.4 show the schematic diagrams of twin-wire systems connected to a single power source with possible welding directions and with an individual power source for GMAW and SAW processes, respectively. When both the wires remain on the same polarity, arcs attract each other and in some ways act like a single elongated arc. In the case of wires arranged in tandem (i.e., one after another), due to mutual interaction, arc at lead wire is subject to

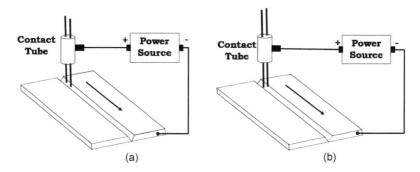

(a)                    (b)

**FIGURE 1.3**
Process diagram for twin-wire welding process with single power source: (a) wires in series—one after another and (b) wires in parallel—one beside another.

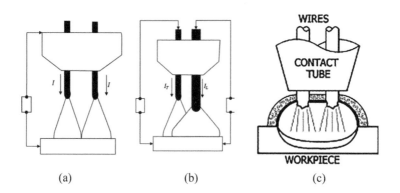

(a)                    (b)                    (c)

**FIGURE 1.4**
Twin-wire system: (a) GMAW multiple electrodes—single power, (b) multiple electrodes—multiple powers (Moinuddin and Sharma 2015), and (c) SAW multiple electrodes—multiple powers (Tusek and Suban 2003). (With kind permission from Elsevier.)

rearward blow, whereas arc at the trail wire experiences forward blow. In this type of arrangement, trail arc controls the final welding shape; the forward deflection allows the use of higher welding speed before encountering weld undercut (Uttrachi 1966).

Twin-wire welding offers higher productivity than its single-wire counterpart. Due to its inherent capability of higher deposition rate and high welding speed range, it can produce welds with 20% less cost in comparison with single-wire welding (Tušek et al. 2005). Deposition rate of this technique is ranging to more than double that of a single electrode. This increase is partially due to increased current density and higher Joule heating of the smaller diameter wire. Twin-wire welding offers the advantage of joint fit-up over other multiple-wire welding systems which use more than one wire with individual power sources (e.g., tandem welding) (Uttrachi 1966). The wire-to-wire axis of the twin-wire torch can be oriented at an angle to the weld line that enables to fill wider grooves. In addition to this, low heat input is another advantage of the twin-wire process. As the power is efficiently used in melting the electrodes, unnecessary dissipation of the heat to the workpiece is prevented. In turn, detrimental effects like residual stress and distortion can be reduced.

The working of GMAW and SAW differs slightly: in GMAW, the arc is exposed to the surrounding, whereas in SAW, the arc is fully covered by a flux, as shown in Figure 1.4, allowing the process to utilize maximum available heat input. In single-power source, twin-wire GMAW/SAW, the current splits in-between the wires in proportion to their cross-section area and is based on welding direction, termed as "parallel and series wire welding", as shown in Figure 1.4a. In the case of individual power sources, two wires are independent of each other and can be controlled separately as shown in Figure 1.4b,c.

### 1.3.2 Welding with Three or More Wires

The twin-wire concept has been extended to three or more wires. The schematic diagrams of a triple-wire system for SAW and GMAW are shown in Figure 1.5a and b, respectively. In the case of wire in series (Figure 1.5), the mid wire is utilized to maintain the stability of the process through reducing the interactions between the arcs. In the case of the triangular arrangement of wires (Figure 1.1), the three pairs of wires interact with each other like twin-wire welding. The net effect of three interactions is a kind of rotary arc directed toward the center that helps to achieve weld penetration.

The wires in four- and six-wire systems can be arranged in several combinations as shown in Figure 1.1. The four- and six-wire welding processes are yet to be established at the shop-floor level. Therefore, developments in multiple-wire technology are dominated by two-wire arrangement systems. In this arrangement, twin-wire is commonly used for overlays, whereas tandem is the most popular version for welding.

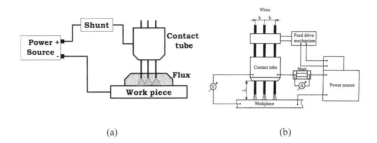

**FIGURE 1.5**
Triple-wire system with a single power source: (a) SAW and (b) GMAW (Tusek 2004). (With kind permission from Elsevier.)

## 1.4 Recent Advancements in Multiple-Wire Technology

The recent advancements in multiple-wire technology are mainly due to advancements in the power source technology. The tandem GMAW, TIME, twin-tungsten TIG welding, varying polarity welding, double-electrode, etc. are some of the recent noteworthy developments.

### 1.4.1 Tandem GMAW

One of the earliest advancements in multiple-wire GMAW is tandem GMAW. Ueyama et al. (2005a–c) studied tandem GMAW, wherein two torches were integrated into one bigger torch and two close parallel arcs were connected to two independent power supplies. This arrangement allows the arc pressure to remain unchanged, and the wire feed speed can be doubled along with the deposition rate (Li and Zhang 2008). A schematic representation of the tandem GMAW welding system is shown in Figure 1.6. Since two wires remain in close proximity, the self-induced magnetic fields strongly influence each other, known as electromagnetic interaction (EMI). Because of the arc interactions, the weld pool oscillates and the arc deflects, which may result in poor weld bead geometry and mechanical properties (Motta and Dutra 2006; Scotti et al. 2006).

The behavior of magnetic fields depends on the polarities supplied to the wires. The possible polarities and the behavior of magnetic fields are shown in Figure 1.7. When two arcs have opposite polarities, as shown in Figure 1.7a, the magnetic forces created by each of the arcs are in the same direction which results in a strong magnetic field in between the wires; in turn, the arcs deflect away from each other. When two arcs have the same polarity, as shown in Figure 1.7b, the magnetic fields oppose each other, causing the arcs to deflect toward each other.

When the arc currents are supplied from different power sources, one with direct current (DC) and the other with alternating current (AC), as shown in

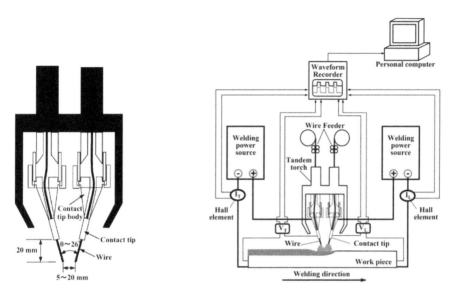

**FIGURE 1.6**
Tandem GMAW system (Ueyama et al. 2005b,c). (With kind permission from Taylor & Francis.)

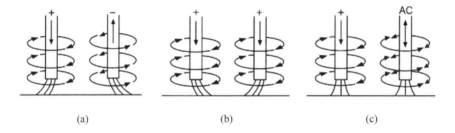

**FIGURE 1.7**
Reactions of magnetic fields when two arcs are in close proximity: (a) arcs are of different polarities: magnetic fields are additive, and the arcs blow outwards; (b) arcs are of the same polarity: magnetic fields are opposite, and the arcs blow inwards; and (c) arcs are in DC and AC: little magnetic blow occurs (Lincoln Electric 1999). (With kind permission from Lincoln.)

Figure 1.7c, the magnetic field of the AC arc completely reverses in each cycle; thus, the DC field is least affected. Therefore, a very little deflection occurs at either arc. In order to reduce the EMIs between the arcs, pulsed and pulsed synergic power sources have been introduced in twin-wire and tandem welding processes. In the synergy mode, unit current pulses detach molten droplets of fixed volume by controlling the parameters necessary for consistent electrode burn-off. The parameters for a given material and wire diameter are pre-programmed into the synergic welding controller such that the pulse parameters are selected automatically (Moinuddin and Sharma 2015).

### 1.4.2 Power Source Synchronization

One of the significant developments in twin-wire welding is power source synchronization. The synchronization of power sources refers to the simultaneous control of power source, as shown in Figure 1.8, such that the mutual interferences of the two wires should be low (Goecke et al. 2001). The mutual interference is minimized by allowing alternate arcing at the two wires. The synchronization is essential as the heavy weld pool wave causes short-circuiting. Improper metal transfer at the rear arc pushes the droplet horizontally into and through the front arc and causes spatter.

The synchronization is achieved by providing pulsed current at lead and trail power sources. The pulsed current waveform of the arc can be designed by selecting appropriate parameters, namely, leading peak current $I_{Lp}$, leading base current $I_{Lb}$, leading peak time $T_{Lp}$, leading base time $T_{Lb}$, trailing peak current $I_{Tp}$, trailing base current $I_{Tb}$, trailing peak time $T_{Tp}$, and trailing base time $T_{Tb}$. The waveform frequency can be calculated by the inverse of the sum of peak time and base time. Considering synchronization, different waveforms can be designed such as the same frequency with 180° phase difference (Figure 1.9a), the same frequency with 0° phase difference (Figure 1.9b), and a different frequency with a phase difference (Figure 1.9c).

Among the three waveform patterns, the first two are simple to design. The second one (i.e., the same frequency and the same phase) leads to fluctuation in electromagnetic force because of large and no electromagnetic force in the peak and base current durations, respectively. The effect of phase difference on arc deflection is shown in Figure 1.10. The deflection with the same frequency and a phase difference of 180° (out-of-phase) leads to stable arcs (Figure 1.10a), while in-phase synchronization allows simultaneous flaring of both the arcs (Figure 1.10b). The arc at the lead wire acts on the cold metal, thus remaining concentrated. The trailing wire excessively flares and causes

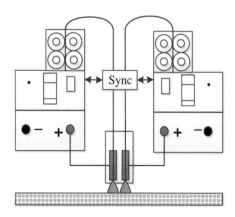

**FIGURE 1.8**
Schematic representation of twin-wire MIG welding with synchronization (Yao et al. 2016).

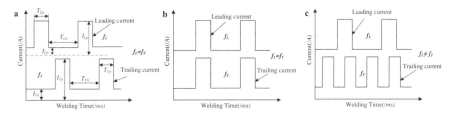

**FIGURE 1.9**
Different current waveforms in twin-wire welding: (a) the same frequency with 180° phase difference, (b) the same frequency with 0° phase difference, and (c) different frequencies with different phase pulses (Yao et al. 2016). (With kind permission from Elsevier.)

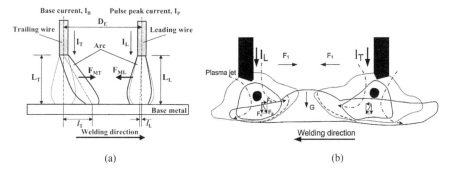

**FIGURE 1.10**
EMIs between the arcs: (a) phase difference 180° (Ueyama et al. 2005b, 521) and (b) phase difference 0° (Chen et al. 2015). (With kind permission from Elsevier and Taylor & Francis.)

arc instability and spatter formation. The 180° phase difference does not fully isolate the arcs as the residual current and magnetic field after extinguishing an arc keep influencing the other arc. Therefore, several investigators used an overlapping waveform (i.e., phase difference less than 180°) wherein the trailing arc is delayed by 0.5–2.5 ms in order to reduce the EMI between the arcs (Schnick et al. 2011; Moinuddin and Sharma 2015; Goecke et al. 2001). The most complex waveform design is different frequencies at lead and trail wires (Figure 1.9c). As the number of controlling parameters doubles, obtaining an optimum wave design becomes very difficult; thus, this design is seldom used.

### 1.4.3 TIME-Twin

TIME-twin by Fronius International GmbH is similar to the tandem welding process. The main feature of this process is TIME which allows achieving a stable electric arc with a desirable metal transfer at both the wires. Unlike conventional GMAW/SAW process, this process uses a patented four-component gas mixture ($Ar/He/CO_2/O_2$) along with a specialized nozzle that controls the metal transfers in all positions and thereby stabilizes the process. This nozzle facilitates larger electrode extension, which, in turn, increases the deposition rates. TIME has a composition of 25%–35% He, 6.7%–8.5% $CO_2$, 0.3%–0.8% $O_2$, and the balance argon for carbon steel; and 52%–60% He, 2.5%–3.4% $CO_2$, 0.1%–0.3% $O_2$, and the balance argon for stainless steel (Gedeon and Catalano 1988; Shona and Bayley 2012). TIME-twin welding torch consists of two contact tubes that are electrically isolated to each other. The synchronization is achieved by master–slave arrangement. The two power sources are interfaced through a local high-speed bus that communicates in between the power sources. One of the power sources acts as a master that triggers the slave power source after a preset time delay (Egerland et al. 2009; Blackman et al. 2002). The power sources can be set with two separate electrode potentials that make them independently controllable. The TIME-twin process can be operated with continuous or pulsed currents at either power source, among which pulse-pulse mode is most preferable. The shape of the pulse plays an important role. The square pulse results in an instantaneous dip in the current from the peak value to the base value that induces arc interruption. Thus, the shape of the pulse is kept trapezoidal with a slow rise and fall, before and after the peak current, respectively. The TIME-twin process allows the use of dissimilar currents at the lead and trail electrodes that facilitate effective utilization of heat. The lead arc that acts on a colder surface is subject to more resistance; thus, a higher current stabilizes the welding arc and yields better mechanical properties (Moinuddin and Sharma 2015).

### 1.4.4 Double-Electrode GMAW and SAW

One of the latest modifications in the multiple-wire welding process is the double-electrode GMAW/SAW process. This novel modification was introduced using a very different mechanism from the existing technologies and implemented in GMAW by the University of Kentucky (Li and Zhang 2008). The double-electrode GMAW (DE-GMAW) has two versions, namely, nonconsumable and consumable, as shown in Figure 1.11a and b, respectively. In the nonconsumable version, a tungsten electrode is used to bypass one part of the melting current in a conventional GMAW process. The total melting current is divided into base metal current and bypass current. The bypass current flows back to the power source through the bypass torch without going through the base metal. Hence, the base metal current is no longer part of the total melting current. The total melting current can be varied to attain an improved deposition rate with reduced heat dissipation to the base metal without any compromise in productivity. If the energy absorbed by the bypass electrode (tungsten electrode) is used in wire melting, it will further improve the productivity. Therefore, another version of the DE-GMAW process has been introduced by replacing the nonconsumable tungsten electrode with a consumable electrode. This is referred to as a consumable DE-GMAW process as shown in Figure 1.11b, where the consumable electrode is fed through a GMAW torch as the bypass electrode.

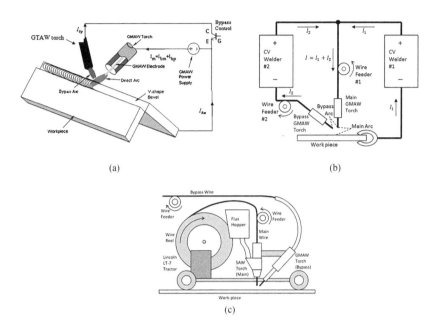

**FIGURE 1.11**
Double-electrode system: (a) nonconsumable DE-GMAW system, (b) consumable DE-GMAW system, and (c) DE-SAW system (Lu et al. 2014). (With kind permission from Elsevier.)

Unlike the nonconsumable DE-GMAW, the consumable DE-GMAW process uses two separate power sources for base metal current and bypass current. If the power sources operate in a constant voltage mode, the flow of total and bypass currents depends on wire feed speed and voltage setting. The main power supply generates the arc between the electrode and workpiece, and the bypass power supply generates an arc between the bypass electrode and the main electrode which closes the loop. Further modification in DE-GMAW is done by altering the polarity between direct current electrode positive (DCEP) and direct current electrode negative (DCEN) at the bypass electrode. The consumable DE-GMAW is extended to SAW to reduce the distortion caused by excessive heat. Figure 1.11c shows a DE-SAW system wherein a bypass torch (GMAW torch) is coupled to the power supply and the tractor carries the torch setup with a wire feeder. A feedback control system with sensors that control the initialization of the bypass arc with delay and wire feed speed to control the base metal current and bypass current (Lu et al. 2014). The stability between the arcs in this process is easier to maintain than the open arc DE-GMAW, thereby achieving increase in productivity with minimum loss of metal.

## 1.4.5 Twin-Wire Indirect Arc Welding

A novel technology in multiple-wire welding is the twin-wire indirect arc (TWIA) welding process introduced to redistribute the energy between

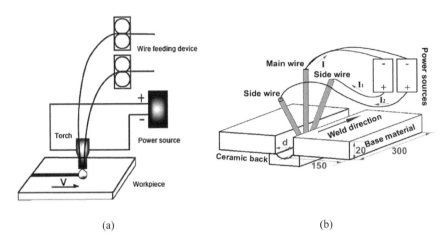

**FIGURE 1.12**

Schematic diagram of indirect arc welding: (a) twin-wire (Shi et al. 2014a,b) and (b) triple wire (Fang et al. 2016). (With kind permission from Elsevier and Springer.)

the wires and workpiece. A schematic representation of TWIA is shown in Figure 1.12a, wherein two electrodes are connected to the positive and negative terminals of the DC power source and the workpiece is independent of the current circuit (i.e., no electrical connection). The arc burns between the wires. The advantage of this process is high melting efficiency with low heat input.

The electromagnetic characteristics of the arcs strongly influence temperature, velocity, and heat flux which are responsible for shaping the arc and weld pool. The possible metal transfer modes are globular–globular transfer, projected–globular transfer, shorting–circuiting transfer, projected–projected transfer, streaming–projected transfer, and streaming–streaming transfer (Zhang et al. 2009). The TWIA is designed especially for the applications that require low penetration (Shi et al. 2014a,b). The indirect arc is extended to triple-wire welding, as shown in Figure 1.12b, wherein three wires are connected to two power sources and the base material is not linked to the power source. The main wire (i.e., middle wire) is connected to the negative terminal, while the side wires are connected to the positive terminal of the power sources. The electric currents $I_1$ and $I_2$ are carried by the side wires, while the main wire carries the current $I$. The current $I$ is the sum of $I_1$ and $I_2$. If $I_1=I_2$, the indirect arcs that establish between each side wire and the main wire remain symmetrical; otherwise, they become nonsymmetrical. The power source modes (DC and pulsed DC) and the contact angle between the main wire and the side wires are used to stabilize the process by reducing the EMI between the arcs.

### 1.4.6 Cable-Type Welding Wire Arc Welding

One of the recent variants of multiple-wire welding is cable-type welding wire (CWW) arc welding (Fang et al. 2017; Chen et al. 2017). The CWW consists

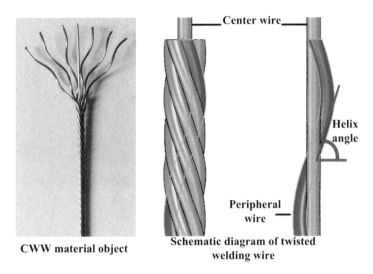

Center wire

Helix angle

Peripheral wire

**CWW material object**

**Schematic diagram of twisted welding wire**

**FIGURE 1.13**
Electrode for CWW (Fang et al. 2017). (With kind permission from Elsevier.)

of a twisted strand of multiple wires surrounding the center wire as shown in Figure 1.13. This variant of multiple-wire is reported to have improvement in deposition rate by 40% and energy saving of 25% compared to $CO_2$ welding using single-wire of the same diameter as that of cable (Fang et al. 2017). The multiple wires ensure continuous feed of the electrode material.

The arcing phenomenon of CWW is very interesting. The rotating arc phenomenon with more than two wires, as discussed earlier, is augmented by the peripheral wire strands. The peripheral wires, due to their helix angle (Figure 1.13), rotate along the inverse wire stranding direction. The arc rotation reduces the arc pressure that produces a shallow but uniform weld pool. The rotating arcs facilitate the rotation of liquid metal in the weld pool that allows faster cooling. The agitation in the weld pool leads to finer and uniform grains with improved hardness.

### 1.4.7 Metal Powered-Assisted Multiple-Wire Welding

Secondary process variants such as the metal powder-assisted SAW are also in use in multiple-wire welding. The metal powder addition before flux is used for surfacing, reducing flux consumption, increasing arc efficiency, and thereby improving penetration and deposition rate. In the twin-wire welding, there are two possible arrangements for metal addition—between the wires and ahead of the welding head, as shown in Figure 1.14a and b, respectively. In the case of three or more wires, the arrangement of metal powder addition should be between the wires as shown in Figure 1.14c. The consistency in the powder flow is critical for a successful implementation

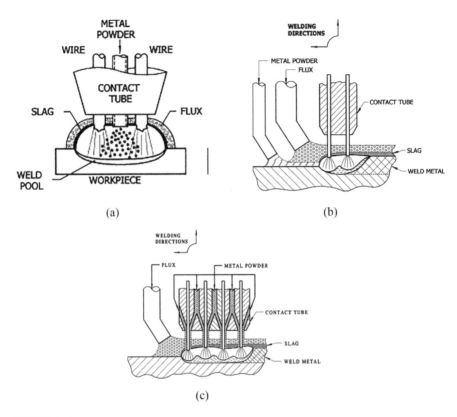

**FIGURE 1.14**
Possible arrangements for metal powder addition: (a) twin-wire: between two wires, (b) double-wire: ahead of welding head, and (c) multiple-wire: between the wires (Tusek and Suban 2003). (With kind permission from Elsevier.)

of the metal power-assisted welding. Inconsistent feeding may result in uneven bead or insufficient penetration (i.e., cladding in the place of welding), thereby lowering the overall melting efficiency of the weld.

## 1.5 Recent Trends in Research in Multiple-Wire Welding

Each process variant discussed in the previous section is unique in its own and implemented to improve deposition and production rate. Major challenges in multiple-wire welding are the control of EMIs between the arcs and improving the stability. Therefore, there is a keen interest in the investigation of improving arc stability. The process parameters such as current, wire diameter, arc length, inclination angle for each wire, and wire extension, which can be controlled and adjusted independently, are used to improve arc stability.

Various methods are proposed to reduce the interaction between the arcs, such as antiphase synchronization between the currents of each wire (i.e., lead wire in peak current, trail wire in the base current, and vice versa) (Motta and Dutra 2006; Moinuddin et al. 2015), distance between the wires (Ueyama et al. 2005a), current–duty ratio (Ueyama et al. 2005b), inclination angle of the wires, wire extension (Ueyama et al. 2005a), dissimilar wire diameter (Sharma et al. 2009a–c), arc length control (Groetelaars et al. 2009), and control of gas mixture (Schnick et al. 2011). Therefore, the measurement of arc stability becomes an imperative step for the multiple-wire welding. The electric and acoustic signals are two most frequently used attributes in analyzing and determining the stability of the arc-welding process using techniques such as cyclograms and probability density distribution (PDD; Suban and Tušek 2003; Moinuddin and Sharma 2015). In recent years, these techniques have been used in real-time welding with feedback systems that check for the stability of metal transfer and defects during welding. The impact of process parameters on arc stability is assessed through various factors such as welding current and voltage signals (Li and Zhang 2007; Luksa 2006; Pires et al. 2007; Mazzaferro and Machado 2009; Feng et al. 2009), arc light (Zhiyong et al. 2011), and noise emitted during the welding (Grad et al. 2004; Pal et al. 2009, 2010). Among these methods, analysis of the welding current and voltage signals is often used for assessing their reliability and simplicity (Di et al. 2000). Suban and Tušek (2003) presented several methods to evaluate arc stability in GMAW like voltage-current cyclogram, PDD of current and voltage, burning and short-circuiting time of an arc, and the probability distribution of the short-circuiting. Simultaneously, advances in the modeling-based investigation have taken place in order to understand the arc stability of the process. These investigations are carried out into modeling and simulation of heat and fluid flow (Schnick et al. 2011; Ding et al. 2013), prediction of deposition rates (Sharma et al. 2009a–c; Kim et al. 2016), flux consumption (Sharma et al. 2008a,b, 2009a–c), weld bead geometry (He et al. 2011, 2012; Sharma et al. 2015; Moinuddin and Sharma 2015), and melting efficiency (Moinuddin and Sharma 2016). Furthermore, parametric studies have been carried out to understand the effects of the process on the structure and properties of welded joints. (Yu et al. 2016; Sharma 2016; Moinuddin et al. 2016).

## 1.6 Conclusions

The multiple-wire welding is known for its high deposition at higher speeds. Over the past decades, many investigators put their concentrated efforts on multiple-wire SAW process and in the recent past on multiple-wire GMAW process. These processes are highly beneficial in the fabrication of pressure vessels, pipeline industries, shipbuilding industries, etc. The twin-wire version of multiple-wire welding is successfully used on the shop floor. Its variants such as tandem, TIME-twin, double-electrode, and indirect arc are currently in use. Other versions such as triple-wire, four-wire, and six-wire are yet to be fully implemented. In recent years, multiple-wire welding is used also in generating additive manufactured metallic components and functional gradient materials. The major concern in multiple-wire welding is the arc instability due to the electromagnetic intersection between the arcs that can be reduced but cannot be completely eliminated. With the help of advanced instrumentation and mathematical tools, control of arc stability and thereby its effects on energy distribution, deposition, melting, weld bead shape, microstructure, and properties, etc. has been assessed, yet several research efforts are essential to fully understand the thermal, mechanical, and metallurgical aspects of multiple-wire welding.

## References

Ashton, T. 1954. Twin wire submerged arc welding. *Welding Journal* 33: 350–55.

Blackman, S. A., D. V. Dorling, and R. Howard. 2002. High-speed tandem GMAW for pipeline welding. *4th International Pipeline Conference*, Calgary, Alberta, Canada: 517–23.

Chandel, R. S. 1990. Deposition characteristics of the twin wire submerged-arc welding process. *Proceedings of the International Offshore Mechanics and Arctic Engineering Symposium*, Houston, TX, USA, 3: 387–90.

Chen, D., M. Chen, and C. Wu. 2015. Effects of phase difference on the behavior of arc and weld pool in tandem P-GMAW. *Journal of Materials Processing Technology* 225: 45–55.

Chen, S., L. Wang, P. Wei, et al. 2016. Sustaining the inter-wire arc in twin-wire indirect arc welding. *Journal of Manufacturing Processes* 21: 69–74.

Chen, Y., C. Fang, Z. Yang, et al. 2017. Cable-type welding wire arc welding. *The International Journal of Advanced Manufacturing Technology* doi: 10.1007/s00170-017-0943-4.

Di, L., S. Yonglun, and Y. Feng. 2000. Online monitoring of weld defects for short-circuit gas metal arc welding based on the self-organizing feature map neural networks. *Proceedings of the IEEE-INNS-ENNS: International Joint Conference on Neural Networks (IJCNN)*, Como, Italy, 5: 239–44.

Ding, X., H. Li, L. Yang, et al. 2013. Numerical simulation of metal transfer process in tandem GMAW. *The International Journal of Advanced Manufacturing Technology* 69(1–4): 107–12.

Egerland, S., G. Hills, and W. Humer. 2009. Using the time twin process to improve quality and reduce cost in high deposition welding of thick section aluminium. *Proceeding of the IIW International Conference on Advances in Welding and Allied Technologies*, Singapore: 9–14.

Eisenhauer, J. 1998. Industry vision workshop on welding. *Sponsored by US DoE and American Welding Society*, Gaithersburg, MD, USA.

Fang, C., Y. Chen, Z. Yang, et al. 2017. Cable-type welding wire submerged arc surfacing. *Journal of Materials Processing Technology* 249: 25–31.

Fang, D., and L. Liu. 2016. Analysis of process parameter effects during narrow-gap triple-wire gas indirect arc welding. *International Journal of Advanced Manufacturing Technology* 88: 2717–25.

Feng, J., H. Zhang, and P. He. 2009. The CMT short-circuiting metal transfer process and its use in thin aluminium sheets welding. *Materials & Design* 30: 1850–52.

Gedeon, S. A., and J. E. Catalano. 1988. Reduction of M1 weld fabrication costs: The effect of weld shielding gas composition. No. MTL-TR-88-49. *Army Lab Command Watertown Ma Material Technology Lab*, Watertown, MA, USA.

Goecke, S., F. Beschichtungstechnik, U. Berlin, J. Hedegård, et al. 2001. Tandem MIG/MAG welding. *A Welding Review Published by Esab* 56: 24–28. www.researchgate.net/publication/266214551_Tandem_MIGMAG_Welding (Accessed December 29, 2017).

Grad, L., J. Grum, I. Polajnar, et al. 2004. Feasibility study of acoustic signals for on-line monitoring in short circuit gas metal arc welding. *International Journal of Machine Tools and Manufacture* 44: 555–61.

Groetelaars, P. J., C. O. de Morais, and A. Scotti. 2009. Influence of the arc length on metal transfer in the single potential double-wire MIG/MAG process. *Welding International* 23: 112–19.

He, K. F., J. G. Wu, X. J. Li, et al. 2011. Prediction model of twin-arc high speed submerged arc weld shape based on improved BP neural network. *Advanced Materials Research* 216: 194–99.

He, K. F., Q. Li, and J. Chen. 2012. Regression analysis of the process parameters effect on weld shape in twin-arc SAW. *Journal of Convergence Information Technology* 7: 84–92.

Hinkel, J. E., and F. W. Frosthoefel. 1976. High current density submerged arc welding with twin electrodes. *Welding Journal* 55: 175–80.

Kim, C., Y. Ahn, K. B. Lee, et al. 2016. High-deposition-rate position welding of Al 5083 alloy for spherical-type liquefied natural gas tank. *Proceedings of the Institution of Mechanical Engineers, Part B: Journal of Engineering Manufacture* 230: 818–24.

Lassaline, E., B. Zajaczkowski, and T. H. North. 1989. Narrow groove twin-wire GMAW of high-strength steel. *Welding Journal* 68: 53–58.

Li, K., and Y. M. Zhang. 2007. Modeling and control of double-electrode gas metal arc welding process. *2nd IEEE Conference on Industrial Electronics and Applications (ICIEA)*, Harbin, China: 495–500.

Li, K. H., and Y. M. Zhang. 2008. Consumable double-electrode GMAW-part 1: The process. *Welding Journal* 87: 11s–17s.

Lincoln Electric. 1999. How to prevent arc blow. www.lincolnelectric.com/en-us/support/welding-how-to/Pages/preventing-arc-blow-detail.aspx (Accessed December 29, 2017).

Lu, Y., S. J. Chen, Y. Shi, et al. 2014. Double-electrode arc welding process: Principle, variants, control and developments. *Journal of Manufacturing Processes* 16: 93–108.

Luksa, K. 2006. Influence of weld imperfection on short circuit GMA welding arc stability. *Journal of Materials Processing Technology* 175: 285–90.

Mandelberg, S. L., and V. E. Lopata. 1966. High-speed twin-electrode submerged welding process. *Welding Production* 2: 25–28.

Mazzaferro, J. A. E., and I. G. Machado. 2009. Study of arc stability in underwater shielded metal arc welding at shallow depths. *Proceedings of the Institution of Mechanical Engineers, Part C: Journal of Mechanical Engineering Science* 223: 699–709.

Michie, K., S. Blackman, and T. E. B. Ogunbiyi. 1999. Twin-wire GMAW: Process characteristics and applications. *Welding Journal* 78: 31–34.

Moinuddin, S. Q., A. Kapil, K. Kohama, et al. 2016. On process–structure–property interconnection in anti-phase synchronized twin-wire GMAW of low carbon steel. *Science and Technology of Welding and Joining* 21: 452–59.

Moinuddin, S. Q., and A. Sharma. 2015. Arc stability and its impact on weld properties and microstructure in anti-phase synchronized synergic-pulsed twin-wire gas metal arc welding. *Materials & Design* 67: 293–302.

Moinuddin, S. Q., and A. Sharma. 2016. Melting efficiency in anti-phase synchronized twin-wire gas metal arc welding. *Proceedings of 10th International Conference on Trends in Welding Research and 9th International Symposium Japan Welding Society (JWS)*, Tokyo, Japan: 562–65.

Motta, M. F., and J. C. Dutra. 2006. Effects of the variables of the double wire MIG/MAG process with insulated potentials on the weld bead geometry. *Welding International* 20: 785–93.

Pal, K., S. Bhattacharya, and S. K. Pal. 2009. Prediction of metal deposition from arc sound and weld temperature signatures in pulsed MIG welding. *The International Journal of Advanced Manufacturing Technology* 45: 1113–30.

Pal, K., S. Bhattacharya, and S. K. Pal. 2010. Investigation on arc sound and metal transfer modes for on-line monitoring in pulsed gas metal arc welding. *Journal of Materials Processing Technology* 210: 1397–10.

Pires, I., L. Quintino, and R. M. Miranda. 2007. Analysis of the influence of shielding gas mixtures on the gas metal arc welding metal transfer modes and fume formation rate. *Materials & Design* 28: 1623–31.

Reis, R. P., D. Souza, and D. F. Filho. 2015. Arc interruptions in tandem pulsed gas metal arc welding. *Journal of Manufacturing Science and Engineering* 137: 011004.

Schnick, M., G. Wilhelm, M. Lohse, et al. 2011. Three-dimensional modelling of arc behaviour and gas shield quality in tandem gas–metal arc welding using anti-phase pulse synchronization. *Journal of Physics D: Applied Physics* 44(18): 185–205.

Scotti, A., C. O. Morais, and L. O. Vilarinho. 2006. The effect of out-of-phase pulsing on metal transfer in twin-wire GMA welding at high current level. *Welding Journal* 85: 225s–30s.

Sharma, A. 2016. A comparative study on mechanical properties of single-and twin-wire welded joints through multi-objective meta-heuristic optimization. *International Journal of Manufacturing Research* 11: 374–93.

Sharma, A., N. Arora, and B. K. Mishra. 2008a. A practical approach towards mathematical modeling of deposition rate during twin-wire submerged arc welding. *The International Journal of Advanced Manufacturing Technology* 36: 463–74.

Sharma, A., N. Arora, and B. K. Mishra. 2008b. Mathematical modeling of flux consumption during twin-wire welding. *The International Journal of Advanced Manufacturing Technology* 38: 1114–24.

Sharma, A., N. Arora, and B. K. Mishra. 2009a. Analysis of flux consumption in twin-wire submerged arc-welding process with unequal wire diameters. *Trends in Welding Research: ASM Proceedings of the International Conference*, Mountain, GA, USA: 626–30.

Sharma, A., N. Arora, and B. K. Mishra. 2009b. Predictive modelling and sensitivity analysis of flux consumption rate in twin-wire submerged arc welding. *Journal of Manufacturing Technology Research* 1: 287–303.

Sharma, A., N. Arora, and B. K. Mishra. 2009c. Statistical modeling of deposition rate in twin-wire submerged arc welding. *Proceedings of the Institution of Mechanical Engineers, Part B: Journal of Engineering Manufacture* 223: 851–63.

Sharma, A., N. Arora, and B. K. Mishra. 2015. Mathematical model of bead profile in high deposition welds. *Journal of Materials Processing Technology* 220: 65–75.

Sharma, A., N. Arora, and S. R. Gupta. 2005a. Simulation of effect of weld variables on thermal cycles during twin wire welding. *Trends in Welding Research: Proceedings of the 7th International Conference ASM International*, Pine Mountain, GA, USA: 71–77.

Sharma, A., N. Arora, and S. R. Gupta. 2005b. Three dimensional transient heat transfer modeling of submerged twin arc welding process. *Proceedings of 2nd International Conference on Welding and Joining Frontiers of Materials Joining*, Tel Aviv, Israel: 222–30.

Shi, C., Y. Zou, Z. Zou, et al. 2014a. Twin-wire indirect arc welding by modeling and experiment. *Journal of Materials Processing Technology* 214: 2292–99.

Shi, C., Y. Zou, Z. Zou, et al. 2014b. Physical characteristics of twin-wire indirect arc plasma. *Vacuum* 107: 41–50.

Shona, R. M., and C. Bayley. 2012. Microstructural heterogeneities in pulsed gas metal arc welds. *ASM Proceedings of the International Conference: Trends in Welding Research*, Mountain, GA, USA.

Suban, M., and J. Tušek. 2003. Methods for the determination of arc stability. *Journal of Materials Processing Technology* 143: 430–37.

Suga, T., S. Nagaoka, T. Nakano, et al. 1997. An investigation into resistance to porosity generation in high-speed horizontal fillet $CO_2$ welding. *Kobelco Technology Review* 20: 37–41.

Tusek, J. 2000. Mathematical modeling of melting rate in twin wire welding. *Journal of Materials Processing Technology* 100: 250–56.

Tusek, J. 2004. Mathematical modelling of melting rate in arc welding with a triple-wire electrode. *Journal of Materials Processing Technology* 146: 415–23.

Tusek, J., and M. Suban. 2003. High-productivity multiple-wire submerged-arc welding and cladding with metal-powder addition. *Journal of Materials Processing Technology* 133: 207–13.

Tušek, J., I. Umek, and B. Bajcer. 2005. Weld-cost saving accomplished by replacing single-wire submerged arc welding with triple-wire welding. *Science and Technology of Welding and Joining* 10: 15–22.

Ueyama, T., T. Ohnawa, K. Yamazaki, et al. 2005a. High-speed welding of steel sheets by the tandem pulsed gas metal arc welding system. *Transactions of JWRI* 34: 11–18.

Ueyama, T., T. Ohnawa, M. Tanaka, et al. 2005b. Occurrence of arc interference and interruption in tandem pulsed GMA welding-study of arc stability in tandem pulsed GMA welding. *Science and Technology of Welding and Joining* 12: 523–29.

Ueyama, T., T. Ohnawa, M. Tanaka, et al. 2005c. Effects of torch configuration and welding current on weld bead formation in high speed tandem pulsed gas metal arc welding of steel sheets. *Science and Technology of Welding and Joining* 10: 750–59.

Uttrachi, G. D. 1966. Multiple electrode system of SAW. *Welding Journal* 57: 15–22.

Yao, P., J. Xue, K. Zhou, et al. 2016. Symmetrical transition waveform control on double-wire MIG welding. *Journal of Materials Processing Technology* 229: 111–20.

Yu, S. F., N. Yan, and Y. Chen. 2016. Inclusions and microstructure of Ce-added weld metal coarse grain heat-affected zone in twin-wire submerged-arc welding. *Journal of Materials Engineering and Performance* 25: 2445–53.

Zhang, S., D. Wu, and M. Cao. 2009. Metal transfer modes of twin-wire indirect arc welding. *Frontiers of Materials Science in China* 3: 426.

Zhiyong, L., T. S. Srivatsan, Z. Hongzhi, et al. 2011. On the use of arc radiation to detect the quality of gas metal arc welds. *Materials and Manufacturing Processes* 26: 933–41.

# 2

# Regulated Metal Deposition (RMD™) Technique for Welding Applications: An Advanced Gas Metal Arc Welding Process

**Subhash Das**
*ITW India Pvt Limited*
*Pandit Deendayal Petroleum University*

**Jaykumar J. Vora**
*Pandit Deendayal Petroleum University*

**Vivek Patel**
*Pandit Deendayal Petroleum University*
*Northwestern Polytechnical University*

## CONTENTS

2.1 Introduction ................................................................................................ 23
2.2 Regulated Metal Deposition—The Process ........................................... 25
    2.2.1 Working Principle ........................................................................... 25
    2.2.2 Applications of RMD Process ....................................................... 28
2.3 Summary and Future Trends .................................................................... 31
References ............................................................................................................ 31

## 2.1 Introduction

In the ever-changing world today, demand for achieving sustainability in welding processes is the prime objective of the industries. As organizations look for ways to improve their growth and profitability and to maintain a competitive edge, there are various solutions and technologies available. Welding plays a substantial role in the on-time completion of the complex fabrication projects both in the pre-fabrication phase generally carried out in the shop and on-site installation. Gas metal arc welding (GMAW) is one such process where a continuous consumable wire is melted and the required

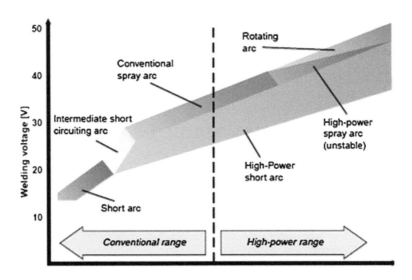

**FIGURE 2.1**
Different modes of droplet transfer in conventional GMAW process [1]. (With kind permission from Springer.)

components are joined. The continuous supply of the wire helps the manu-facturers to achieve increased productivity as well as to attain the required quality. GMAW process incorporates different metal-transfer modes which are basically a function of the welding parameters such as welding voltage (V) and welding current (A), as shown in Figure 2.1. However, due to the dynamic nature of the arc welding process, control of these parameters is always a difficult task for the conventional power sources. This accuracy can be enhanced by using digital and electronic controls in power sources [1].

A modification in the power sources and the involvement of computer tech-nology made the designing of the intended waveform and subsequently the metal transfer and deposition possible particularly after 1990s [2]. Scientific invention brought an impeccable variation to the welding processes, and particularly for GMAW process, the transfer modes were categories as a nat-ural metal transfer, controlled transfer, and extended operating techniques as shown in Table 2.1 [3]. Furthermore, as compared to an earlier approach wherein the torch motion was kept constant and an adjustment of the wire speed was done, the new advances integrated both the power source and wire feeder to reach an augmented molten metal-transfer mode in GMAW. This approach is termed as "mechanically assisted droplet transfer" and is widely applied in controlled short-circuiting mode by retracing the wire.

The key advances in power source regulation have been done for the short-circuiting process. The developments of new arc welding processes were aimed to overcome the restrictions of traditional short arc waveform by enabling new shapes of the arc curve. These power sources benefited from

**TABLE 2.1**

Classification of the Controlled Transfer Mode [1]

| Metal-Transfer Modes | | Welding Process |
|---|---|---|
| Controlled spray | Pulsed transfer | GMAW using variable frequency pulse and drop spray transfer |
| Controlled short-circuiting | Current controlled dip transfer | GMAW using current controlled power source |
| | Controlled wire-feed short-circuit mode | GMAW with wire-feed oscillation |

the enhancement in digital control as well as advanced software, which enabled effective monitoring of almost every aspect of the welding arc. In addition to this, the short-circuiting mode is predictable, can be set at a specific time, and the molten material transfer can be handled easily to minimize the spatter from molten weld puddle [1].

Regulated metal deposition (RMD) technique is one such metal-transfer technique introduced by Miller Electric Company in 2004 and is a variant of the GMAW process which employs the short-circuit principle. The RMD technique typically is founded on the application of advanced software for modified short-circuit transfer GMAW welding that screens the welding current in every step of the short-circuiting phase. The wave profile typically depends on the material composition and thickness to be welded. Several works have been reported regarding the application of RMD technique on different steels. This chapter aims to introduce the basics of the technique and provide a summary of the application of RMD technique on different steels and its beneficial outcomes. To the best of authors' knowledge, an extreme shortage of the scientific literature on the subject exists and the majority of the collected literature are reported by the manufacturer (Miller Electric Company).

## 2.2 Regulated Metal Deposition—The Process

### 2.2.1 Working Principle

The RMD technique anticipates and controls the short-circuit phase and selectively reduces the welding current in order to create a steady metal transfer. An accurately controlled metal transfer delivers an undeviating droplet deposition, which makes it easy for the welder to control the puddle [4]. In order to understand the precise way in which the short circuit is controlled, a typical RMD cycle is divided into seven different phases as shown in Figure 2.2.

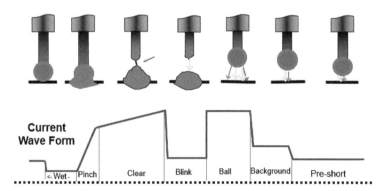

**FIGURE 2.2**
Typical RMD cycle.

In the RMD process, through the opening phase of the metal-transfer cycle (Pre-short), the droplet is formed initially on the melting tip of the electrode wire. Subsequent to this, during the "Wet phase", welding current is reduced owing to which as the wire-feed rate which is similar to the melting rate of the filler metal also increases. Thus, the arc length is reduced leading to the start of a short-circuit condition. During the next phase, i.e., "short-circuit" phase, the welding current waveform is distributed into two phases. In the initial phase, known as "Pinch phase", the current rapidly increases once the droplet comes in contact of the weld pool. In the subsequent phase, known as "Clear phase", the current continues to rise, but at a modest rate, until the power source control detects the end of the short-circuit phase. Once the droplet is detached, the current again reduces during a brief period known as "Blink phase", so that the electric arc could smoothly reignite, which in turn avoids spattering too. As a result, the welding current is increased again in the "Ball phase" promoting the generation of a new metal droplet on the tip of the wire. During the subsequent "Background" and "Preshort" phases, the welding current is gradually reduced (in step shape) allowing new contact between the molten tip and weld pool, which guarantees better weld-pool stability. The cycle thus continues and the molten metal is deposited in a calm manner by integration of these seven phases [5].

The RMD software program, synergically working with an inverter-based welding system and closed-loop feedback, closely monitors and controls the electrode current at a speed of up to 50 μs (50 millionths of a second). In addition to this, the software also precisely regulates the essential welding speed and gas combination for a definite wire diameter [6,7].

A modified short-circuit GMAW process offers numerous benefits that can help improve productivity, efficiency, and quality. The idea behind developing the technique was conceived from the fact that when a pebble is dropped into a lake the ripples created in the lake water appear almost unnoticeable.

However, when a brick is dropped into the lake it produces a big splash which is harder to control. The finer, more consistent droplets produced in a modified short-circuit GMAW process are like the pebble, where it's easier to control the impact, while the transfer produced in a traditional short-circuit GMAW process is more like the brick, creating sidewall splashing of the weld puddle. In a traditional GMAW process, this "splash" can lead to issues of the cold lap in the finished weld. Thus, the more consistent molten pool of the RMD process helps to control the metal transfer in a more effective way [8].

This modified short-circuit process provides several practical benefits in real-time fabrication. Some of these advantages are listed the following [4,6–13].

1. The smooth metal transfer achieved with the RMD technique compensates for a high–low misalignment between pipe sections. This ensures that the time taken for the prior fit-up can be reduced greatly.

2. In pipe welding, the smooth and calm metal transfer generates reliable root reinforcement on the inner side of the pipe which is otherwise a challenge due to limited accessibility.

3. The shielding gas coming out from the welding gun remains relatively undisturbed due to the controlled transfer. Thus, enough shielding gas gets strapped through the root opening which prevents the sugaring (oxidation) on the rear side of the weld. This is an important advantage to count because large-diameter pipes take a long time to purge. Using this technique, the backing gas can be eliminated while welding, which significantly improves productivity. The lesser usage of shielding gas welding also guarantees cost saving as these gases are quite costly.

4. The RMD technique sustains a steady arc length regardless of electrode stick-out. This ability is particularly beneficial when operators do not hold a constant stick-out, to get a good view of the weld puddle. In addition, because the weld puddle is easier to read and manipulate, new welding operators can be trained faster to produce quality welds. It also provides more comfort to the welder.

5. The RMD technique can generate a root pass weld with a 3–6 mm throat. The amount of metal deposited during root pass would be sufficient to support the heat input requirements of the subsequent fill up passes for GMAW and FCAW process, whereas for TIG welding applications, the possibility of removal of the hot pass can also be considered.

6. The weld puddle created by RMD technique freezes faster than in traditional GMAW processes, which allows the welding operator to run a higher wire speed, thus improving the deposition of each pass.

**TABLE 2.2**

RMD Process in Comparison with Conventional Processes

| Process | Travel Speed (in./min) | Training Time | Slag Clean Up | Repositioning/ Starts and Stops |
|---------|------------------------|---------------|---------------|----------------------------------|
| RMD | 6–12 | Low | No | Infrequent |
| TIG | 3–5 | High | No | Frequent |
| SMAW | 3–8 | Moderate | Yes | Frequent |

7. The lower heat input with RMD process helps improving the mechanical properties of the finished weld.
8. The smooth droplet transfer and stable weld puddle result in reduced rework, leading to a decrease in the costs, meeting quality requirements, and improving productivity to meet tighter deadlines. The reduced sidewall splashing of this process also results in reduced post-weld grinding.

The RMD process is quite better as compared to the conventional process, as shown in Table 2.2. Thus, the RMD technique can ensure fabricators a productive and sound welding technique to work upon.

## 2.2.2 Applications of RMD Process

To realize these advantages of the RMD process over the conventional process, the fabricators considered the RMD process a candidate process for root pass as well as pipe to pipe joint applications; however, limited scientific literature has been published in the subject. RMD technique was reported to be used on SS 316L tubes with a diameter of 0.5–34 in. in the root pass primarily. RMD process created a root pass weld with a throat of 1/8–1/4 in. which ensured the higher deposition, thus eliminating the hot pass in a certain configuration. It was observed from the study that one of the biggest challenges in pipe welding occurs when two section of pipes are misaligned, also known as the high/low situation. With the conventional processes, the situation would require either a better fit-up or a repair after welding. However, with the RMD technique, up to 1/8 in. misaligned pipes can be welded without any rework as this process can pick up both the edges and can fuse them together as well (see Figure 2.3). One of the important observations reported from the study was the soundness with which the process operates as compared to erratic conventional MIG welding process. A typical crackle sound occurred when the process has some problem. This feature gives the welder an added benefit to identify and rectify in process problem resulting in more reliable final weld [9]. Similarly, Cuhel [4] reported that the root pass by RMD technique was found to be flat with no evident convexity or concavity and was thicker than traditional GMAW root, as shown in Figure 2.4.

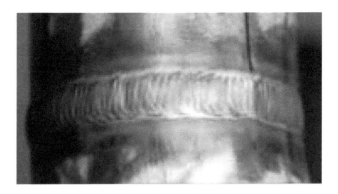

**FIGURE 2.3**
SS tubes welded with RMD.

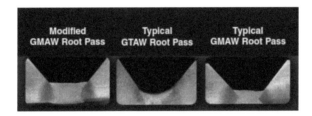

**FIGURE 2.4**
Root pass with RMD.

In a time study analysis conducted by Miller, it was found that RMD technique demonstrated 22% improvement over manual TIG welding. This was also largely due to the fact that majority of the welds carried out were single pass full penetration weld wherein a precise control over the metal transfer and molten pool control was involved. The real benefit of using RMD technique was on SS. With the RMD technique, an X-ray quality weld was achieved without a back purging as shown in Figure 2.5 indicating good penetration without oxidation. This enabled the process to be more productive and resulted in a generous saving in the cost of fabrication of the components [10].

A study reported by Joe Ryan from Miller Electric Mfg. Co. [11] revealed the application of RMD technique on the carbon steel pipes of schedule 80 having a diameter ranging from 6 to 16 in. It was conceived from the study that the traditional GMAW process when applied to the root pass gave problems such as blowing hole and excessive spatter. However, with the RMD technique, the root pass was deposited (see Figure 2.6) with a greater control and robustness giving the process an edge over conventional welds. Another study reported was the targeted application of the RMD technique for the pressure vessel joints. The main challenge in welding of the joints for pressure vessel was the 100% radiography clearance of root pass joints. If there

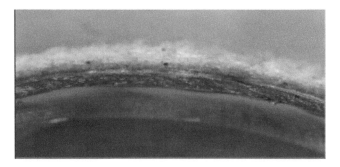

**FIGURE 2.5**
Back side of SS weld deposited with RMD without no backing gas showing good penetration.

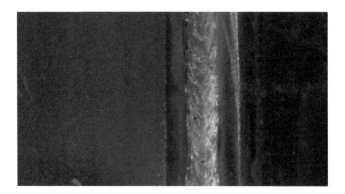

**FIGURE 2.6**
Root pass weld deposited by RMD process.

is a flaw in this root pass, the entire thickness of the joint (which is generally 4 in. thick) has to be gouged and redone. This caused a major setback for the company in terms of productivity. RMD technique was experimented instead of conventional short arc GMAW process and it was observed that the root joints showed better soundness and integrity to even pass the X-ray test reducing the rework, hence increasing the productivity [12].

As stated by manufacturers, the RMD technique employs a different waveform divided across seven different phases. However, limited study is carried out on analyzing the effect of these waveform characteristics on the weld-bead morphological features. Costa and Vilarinho [5] studied the influence of waveform parameters on the current and bead geometry. The study was analyzed by joining of 5.5-mm-thick pipes with a diameter of 63.5 mm in single-pass, by welding in both upward and downward directions. The obvious output from the study was that the internal discontinuities, specifically porosity, inclusion, lack of fusion, and cracking, were not present in the welded joint. The authors also checked the possibility of application of weaving technique and the results indicated that it significantly influenced the penetration reduction. This was because the increase in the weld-bead width favored the reduction in the root reinforcement. Thus, it is assumed that an increase in wire-feed speed led to an increase in penetration and a root-reinforcement, and a decrease in face reinforcement.

## 2.3 Summary and Future Trends

This chapter presented the working principle, several successful applications of the RMD technique, and advantages of the process which are reported by the end users as well as the manufacturers. The technique can be considered as fairly matured with its application on widely used base metals such as SS, carbon steel, etc. However, the literature available in the open forum for the technique is quite less. Thus, it is assumed that the way forward for using the technique would be to use a metal-cored filler wire in the root pass welding. This would be fairly similar to the metal-cored arc welding method [14] used for welding but in a controlled short-circuiting mode. The use of these wires would allow the welding to be done at high duty cycles, faster travel speeds, and lower welding fumes, and hence would be cost-effective. An attempt for the same was reported wherein RMD with metal-cored wires helped significantly to increase travel speeds (compared to TIG), bridge gaps and eliminate lack of fusion or lack of penetration by providing a thicker root pass, reduce spatter (compared to conventional GMAW processes) on root passes, and achieve a better weld compared to a solid wire [13].

## References

1. P. Kah, R. Suoranta, J. Martikainen, "Advanced gas metal arc welding processes", *The International Journal of Advanced Manufacturing Technology*, 67 (2013) 655–674.
2. K. Weman, "Modern MIG welding power sources", *ESAB Welding Equipment AB, Laxa, Sweden, Svetsaren*, 1 (1999) 77–79.
3. W. Lucas, D. Iordachescu, V. Ponomarev, "Classification of metal transfer modes in GMAW", IIW Doc, (2005).
4. J. Cuhel, "Modified GMAW for root passes", *TPJ-The Tube & Pipe Journal*, (2008).
5. L.O. Vilarinho, T.F. Costa, "Influence of process parameters during the pipe welding of low-carbon steel using RMD (regulated metal deposition) process", *Proceedings of COBEM 2011, 21st Brazilian Congress of Mechanical Engineering*, Natal, Brazil, 2011.

6. Miller Electric Mfg. Co. (2004), "Software-driven RMD™ process overcomes short circuit MIG limitations" www.millerwelds.com/about/news_releases/2004_archive/articles82.html. Accessed Jul 2018.

7. Miller Electric Mfg. Co. (2003), "Miller introduces Axcess multi MIG systems—most significant welding invention since the inverter" www.millerwelds.com/about/news_releases/2005/articles140.html. Accessed Jul 2018.

8. Miller Electric Mfg. Co. "Mechanical contractor increases pipe welding productivity up to 500%".

9. Miller Electric Mfg. Co. "Shinn mechanical uses PipeWorx Welding System to increase pipe fabrication quality and productivity" www.millerwelds.com/resources/article-library/shinn-mechanical-uses-pipeworx-welding-system-to-increase-pipe-fabrication-quality-and-productivity. Accessed Jul 2018.

10. Miller Electric Mfg. Co. "Graham corporation smashes reduced rework objectives with help from Miller's PipeWorx Welding Systems" www.millerwelds.com/resources/article-library/graham-corporation-smashes-reduced-rework-objectives-with-help-from-millers-pipeworx-welding-systems. Accessed Jul 2018.

11. Miller Electric Mfg. Co. "Mechanical contractor diversifies and redefines business to meet the demand of the mid-Atlantic steel, oil and gas industries" www.millerwelds.com/resources/article-library/mechanical-contractor-diversifies-and-redefines-business-to-meet-the-demand-of-the-midatlantic-steel-oil-and-gas-industries. Accessed Jul 2018.

12. Miller Electric Mfg. Co. "The pride of Big Prairie, Ohio" www.millerwelds.com/resources/article-library/the-pride-of-big-prairie-ohio. Accessed Jul 2018.

13. Miller Electric Mfg. Co. "RMD short-circuit metal transfer, pulsed MIG processes with metal-cored wires improve pipe fabrication for Swartfager Welding, Inc" www.millerwelds.com/resources/article-library/rmd-and-pulsed-mig-processes-with-metalcored-wires-improve-pipe-fabrication-for-swartfager-welding-inc. Accessed Jul 2018.

14. M. Kazasidis, D. Pantelis, "The effect of the heat input energy on the tensile properties of the AH-40 fatigue crack arrester steel, welded by the use of the robotic metal-cored arc welding technique", *The International Journal of Advanced Manufacturing Technology*, 93 (2017) 3967–3980.

# 3

# *Friction Stir Welding and Its Variants*

**Vivek Patel**
*Northwestern Polytechnical University*
*Pandit Deendayal Petroleum University*

**Wenya Li**
*Northwestern Polytechnical University*

**Devang Sejani**
*Technische Universität Darmstadt*

## CONTENTS

3.1 Introduction to Friction Stir Welding (FSW) ..........................................34
   3.1.1 Basic Working Principle of FSW ....................................................34
   3.1.2 Advantages and Limitations............................................................35
   3.1.3 FSW Variants ......................................................................................37
3.2 Assisted Friction Stir Welding (A-FSW) ................................................37
   3.2.1 Plasma—FSW (PFSW) ......................................................................38
   3.2.2 Gas Torch—FSW (GFSW) ................................................................38
   3.2.3 Laser—FSW (LFSW) ..........................................................................39
   3.2.4 Gas Tungsten Arc Welding—FSW (GTAFSW) ............................39
   3.2.5 Electrically Assisted FSW (EAFSW) ..............................................39
   3.2.6 Induction Heating Tool-Assisted FSW (i-FSW) ..........................40
   3.2.7 Ultrasonic Energy-Assisted FSW (UAFSW)................................41
3.3 Friction Stir Spot Welding (FSSW) ........................................................42
3.4 Stationary Shoulder Friction Stir Welding (SSFSW)............................44
3.5 Bobbin Tool or Self-Reacting Friction Stir Welding (SRFSW) ..............45
3.6 Friction Stir Channeling (FSC)................................................................46
3.7 Friction Stir Processing (FSP)..................................................................46
   3.7.1 Grain Refinement................................................................................47
   3.7.2 Superplasticity....................................................................................48
   3.7.3 Surface and Bulk Composite Manufacturing ..............................48
   3.7.4 Casting and Fusion Weld Repair....................................................49
3.8 Summary......................................................................................................49
Acknowledgments ..............................................................................................49
References...............................................................................................................50

## 3.1 Introduction to Friction Stir Welding (FSW)

Welding is one of the most commonly used fabrication techniques. There are numerous welding processes used for different applications such as arc welding, gas welding, plasma welding, laser welding, electron beam welding, friction welding, etc. Friction stir welding (FSW) was invented at The Welding Institute (TWI) by Thomas et al. (1991). FSW—a kind of friction welding—is a hot-shear joining process (Nandan, Debroy, and Bhadeshia 2008).

### 3.1.1 Basic Working Principle of FSW

In the FSW process, a nonconsumable rotating tool (with a shoulder and pin/probe) is inserted and traversed along the joint surfaces of two rigidly clamped base metals, as schematically presented in Figure 3.1. The shoulder makes a gentle contact with the top surface, generates frictional heat by the rubbing, and thus softens the material to weld. The plastic flow of the material occurs as the tool covers the welding regime. The FSW process consists of the following three different steps: plunge and dwell, traverse, and retract (Mishra, De, and Kumar 2014). In the initial step, the rotation tool inserts into the joint area of base metals (with or without backing plate). The tool shoulder makes an initial contact with the base metals, which generates heat due

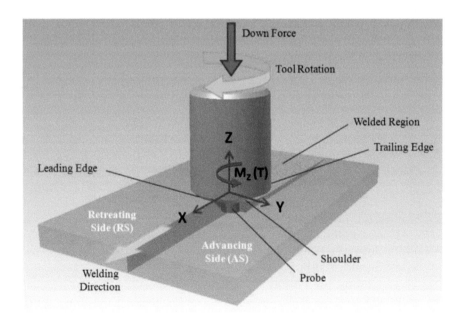

**FIGURE 3.1**
FSW process (Gibson et al. 2014). (With kind permission of Elsevier.)

to rubbing action between tool shoulder and base metal. The tool is made to move forward in the welding direction when the heat generation is enough to displace the material plastically in the form of joint. The traverse advancement of the tool causes the material to flow along the joint line and produces the weld. The traverse motion of the tool is continued until the weld length is obtained and later the tool is retrieved from the joint. A particular nomenclature is given for the two sides of the asymmetric weld (see Figure 3.1): the first is advancing side where the tool travel vector is parallel to the tool rotation direction, and the other is opposite side where the tool traverse and tool rotation direction vectors are antiparallel.

Owning to a solid-state nature of the joining process, FSW has evolved as the technique of choice for lighter and softer materials such as aluminum, copper, and magnesium. Unlike all other conventional welding processes, FSW process also involves a complex interaction of various thermomechanical processes. These interactions result in a variation of heating and cooling rates, plastic deformation, material flow, dynamic recrystallization, and mechanical properties of the joint (Nandan, Debroy, and Bhadeshia 2008). A typical cross-section of the friction stir-welded Al alloy along with temperature and hardness distribution is presented in Figure 3.2. The cross-section consists of four different zones, namely, parent metal, heat-affected zone (HAZ), thermomechanically affected zone (TMAZ), and stir zone (SZ) or weld nugget. The HAZ is similar to the conventional welding process, although the peak temperatures are lower than the solidus temperature and diffusive heat source. Since there is no melting involved, the HAZ properties and microstructure can be somewhat different from the fusion welding process. The SZ or nugget region that appears as "onion rings" is the one which undergoes the most severe plastic deformation during the welding. The onion rings are formed because of the material deposition from front to the back by the tool probe. The TMAZ region lies between SZ and HAZ. The microstructure of the TMAZ is same as that of parent metal but often in a deformed manner. The SZ underwent to the peak temperature, which has dissolved the strengthening precipitates and resulted in a reduction in hardness in the SZ, as shown in Figure 3.2.

### 3.1.2 Advantages and Limitations

Despite being a solid-state welding process, FSW exhibits several advantages and limitations over its conventional fusion counterparts. FSW is mostly done in a single-pass fashion with partial or full joint penetration. Depending on the specimen thickness, workpiece clamping, and process parameters, there can be minimal distortion, which is much less than the fusion welding processes for the same specimens. The mechanical properties exhibited by the weld nugget are superior to those of fusion weld. Moreover, the problems of the fumes spatter and shielding of the weld pool are avoided in FSW. High weld speeds can also be achieved with FSW compared to fusion

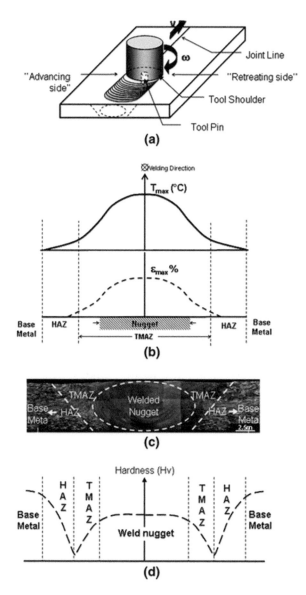

**FIGURE 3.2**
Basics of the FSW process: (a) schematic description of the welding process, (b) maximum temperature and strains associated to the welding process, (c) metallography of a FSW joint between two AA 6056 plates and identification of the different zones, and (d) typical hardness evolution (Gallais et al. 2007). (With kind permission of Springer Nature.)

welding processes, but it is not good enough when compared with laser or electron beam welding processes. As FSW is a solid-state process, common weld defects such as porosity and cracks are absent. FSW possess greater repeatability of the weld in production with lower power output (AWS 2007).

FSW process is used for joining a variety of metals including shiny materials such as steel, aluminum, and copper which are difficult to weld using an autogenous process such as laser welding. Moreover, to join dissimilar materials, the FSW is the best suitable process as it minimizes the formation of intermetallics.

One of the major limitations of the FSW process is that it requires rigid clamping of the base metal due to the large compressive force applied during welding. In addition, to incorporate more than one welding position to fabricate a complex component is nearly impossible with the FSW process. Additionally, the exit pinhole or keyhole on the base metal affects the properties of the components. The initial setup cost of this process is higher than that of fusion welding process (AWS 2007). To weld hard alloys like steel and Ni, the tooling system becomes costly.

### 3.1.3 FSW Variants

FSW is a modern-day joining process, which avoids the use of filler material and joins the material below solidus temperature. For most of the applications where the weight of the component is as an issue, FSW process suits the best. More innovations regarding FSW process has led to the development of its variants. Some of the variants of FSW are as follows:

- Hybrid or Assisted Friction Stir Welding (AFSW)
- Friction Stir Spot Welding (FSSW)
- Stationary Shoulder Friction Stir Welding (SSFSW)
- Bobbing tool or Self Reacting Friction Stir Welding (SRFSW)
- Friction Stir Channeling (FSC)
- Friction Stir Processing (FSP)

All these aforementioned variants of FSW are briefly discussed in the following sections.

## 3.2 Assisted Friction Stir Welding (A-FSW)

Joining of dissimilar materials with FSW process is a bit complex compared to similar material combinations, as both the materials exhibit distinct thermophysical properties. In addition, the formation of intermetallic plays a crucial role when the mechanical properties of the joint are considered. For example, various parameters such as thermal conductivity and material hardness play a major role while joining dissimilar combinations such as aluminum to copper, aluminum to magnesium, aluminum to titanium, and aluminum to steel by FSW. To incorporate these additional parameters, an additional welding process parameter is incorporated into the FSW process. This additional process is used to heat up the harder material so that the heat imposed by FSW process is well enough to weld. This addition of an external heat input source for joining two materials with conventional FSW process is termed as assisted friction stir welding (A-FSW). In A-FSW of dissimilar, the additional heat input source is placed with an offset with the bonding line toward the harder material. The preheated harder material easily flows plastically when the rotating tool comes into contact. In some cases, the additional heat source is also placed behind the conventional FSW process to reduce the cooling rate of the weld nugget (Choi et al. 2011).

Researchers have conducted investigations by using different additional heat input sources. The different AFSW techniques until knowledge are as follows:

- Plasma—FSW (PFSW) (Yaduwanshi, Bag, and Pal 2016)
- Gas Torch—FSW (GFSW) (Choi et al. 2011)
- Laser—FSW (LFSW) (Chang et al. 2011)
- Gas Tungsten Arc Welding—FSW (GTAFSW) (Bang et al. 2012)
- Electrically Assisted FSW (EAFSW) (Ferrando 2008)
- Induction heating tool-assisted FSW (i-FSW) (Vijendra and Sharma 2015; Padhy, Wu, and Gao 2015)
- Ultrasonic energy-assisted FSW (UAFSW) (Padhy, Wu, and Gao 2015)

### 3.2.1 Plasma—FSW (PFSW)

Yaduwanshi, Bag, and Pal (2016) used PFSW for joining aluminum and copper. To overcome the material flow during the dissimilar joining of aluminum and copper, the copper side is heated up to 500–600 K using a plasma torch. The schematic representation of PFSW process is shown in Figure 3.3.

### 3.2.2 Gas Torch—FSW (GFSW)

In the GFSW process, a gas torch is used behind the conventional FSW tooling system. The main objective behind using gas torch is to reduce the cooling rate of the weld nugget. More time is given to the weld nugget, which results in the formation of the more stable phases for a given alloy system.

Choi et al. (2011) demonstrated the GFSW process for welding high-carbon steel. FSWed high-carbon steel specimens resulted in the higher amount of martensite formation in the nugget. In contrast, using gas torch behind the FSW tooling system offers a reduced cooling rate of the nugget region, which results in the lower amount of martensite.

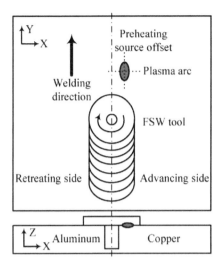

**FIGURE 3.3**
Schematic diagram of plasma-assisted friction stir welding for dissimilar joint (Yaduwanshi, Bag, and Pal 2016). (With kind permission of Elsevier.)

### 3.2.3 Laser—FSW (LFSW)

LFSW is the most widely used hybrid FSW technique. A laser source (most commonly Nd: YAG fiber optic laser, diode laser, and $CO_2$ laser) is used to preheat the workpiece. The preheated workpiece requires a lesser amount of mechanical force by rotating tool of the FSW process. Hence, the welding speed can be increased largely. The general schematic of the LFSW process is shown in Figure 3.4. While joining Al6061-T6 with AZ31 alloy plate using LFSW process, the formation of $Al_{12}Mg_{17}$ brittle intermetallic phase is reduced and more of less brittle Ni phase were formed (Chang et al. 2011).

### 3.2.4 Gas Tungsten Arc Welding—FSW (GTAFSW)

When a tungsten inert gas arc is integrated with conventional FSW process, it is termed as GTAFSW. Tungsten arc is used to preheat the harder material and ease the plastic flow of the material during FSW. Bang et al. (2012) demonstrated GTAFSW for joining Al6061-T6 with stainless steel 304. This study of hybrid FSW has reported the tensile strength of the weld by approximately 93% of the Al6061-T6 alloy.

### 3.2.5 Electrically Assisted FSW (EAFSW)

In the EAFSW process, an additional electric current is applied with the conventional FSW process (Ferrando 2008). This electrical current accelerates the FSW process by providing resistance heating or Joule heating effect and the

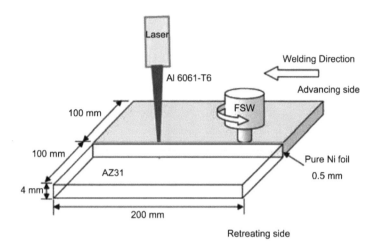

**FIGURE 3.4**
Schematics of the laser-FSW hybrid process (Chang et al. 2011). (With kind permission of Elsevier.)

electroplastic effect (Padhy, Wu, and Gao 2015). The resistance heating supports the preheating of the workpiece and electroplastic effect causes the softening of the material. The electrical current exerted on the workpiece is through the FSW tool. In electroplastic phenomenon, traveling electrons reduces materials resistance toward plastic deformation without a substantial increase in temperature. The EAFSW setup consists of an FSW machine and an electrical circuit to support the FSW process. The process can further distinguish in two ways based on the arrangement of the electrical circuit. In the first category, the circuit is formed by supplying current through the FSW tool, as shown in Figure 3.5. In the second category, the electrical circuit can be formed without accounting any of the machine components of the FSW process.

### 3.2.6 Induction Heating Tool-Assisted FSW (i-FSW)

In induction heating, an induction coil is placed around the electrically conductive material. This induction coil alters its magnetic field and causes the heating of the workpiece. The resistive heating reduces the resistance of the material toward plastic deformation and eases the plastic flow of the material (Padhy, Wu, and Gao 2015). Similar to EAFSW, the i-FSW process can be distinguished into two different ways based on the position of the induction coil. The induction coil can be placed surrounding the tool, as represented in Figure 3.6. It can also be placed ahead of the tool. Placing it surrounding the tool causes heating up of the tool and then transferring it to the base material while placing it ahead of the tool directly heats up the base material.

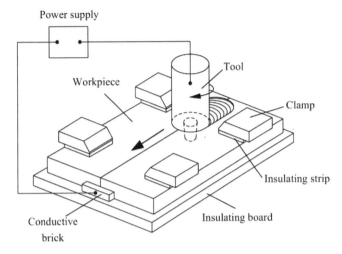

**FIGURE 3.5**
Schematic diagram of electrical-current-aided friction stir welding (Luo, Chen, and Fu 2014). (With kind permission of Elsevier.)

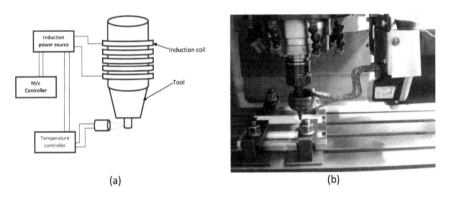

**FIGURE 3.6**
i-FSW: (a) schematic and (b) experimental setup (Vijendra and Sharma 2015). (With kind permission of Elsevier.)

## 3.2.7 Ultrasonic Energy-Assisted FSW (UAFSW)

When high-frequency vibrations are superimposed on a static load, the material can be remarkably softened without significant heating. Ultrasonic waves stimulate blocked dislocations hardened under ordinary deformation and reduce stress for further plastic deformation (Padhy, Wu, and Gao 2015). In the UAFSW process, an ultrasonic transducer is placed directly ahead of the FSW welding head, as shown in Figure 3.7. The high-frequency transducer softens the material ahead of the bonding and high-speed rotating tool stirs and forms a bond between the joint interfaces.

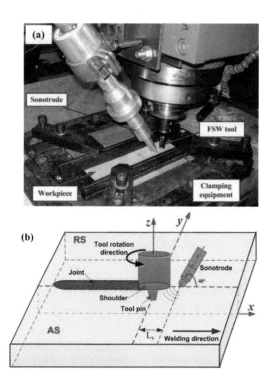

**FIGURE 3.7**
The system of ultrasonic vibration enhanced friction stir welding process: (a) experimental
system and (b) schematic of UVeFSW process (Shi, Wu, and Liu 2015). (With kind permission
of Elsevier.)

## 3.3 Friction Stir Spot Welding (FSSW)

The demand for using lightweight materials by automobile and aerospace
industries is increasing day by day because of the increase in strictness of
the pollution control norms. For spot welding, conventional techniques such
as resistance spot welding (RSW), laser spot welding (LSW), and riveting
are being widely used in automobile and aerospace. Besides its vast usage,
conventional RSW process has many backlogs such as tool degrading, high
power consumption, porosity defect, high heat distortion, and poor joint
strength (Yang, Fu, and Li 2014). Hence, demand for new convenient spot
welding process is very high in today's competitive market.

Mazda Motor Corporation invented FSSW as a variant of FSW in 1993 for
solid-state spot joining of metal sheets. The FSSW process is similar to conventional FSW process without traverse motion of the rotating tool (Patel,
Sejani, et al. 2016). The conventional FSSW process consists of three different
steps, namely, plunging, stirring, and retracting, as illustrated in Figure 3.8.

The only backlog of the conventional FSSW process is the exit keyhole at the end of the process. The properties of the components are adversely affected because the exit pinhole is almost of the entire lap configuration cross-section.

To overcome the limitation of the conventional FSSW, several variants of the FSSW were discovered such as refill-FSSW, swing FSSW, and stitch-FSSW. In the refill-FSSW process, the material displaced by the probe is stored in the sleeve and is used to fill up the exit pinhole, as shown in Figure 3.9. The tool of refill-FSSW consists of three different parts, namely, pin, sleeve, and clamp. The function of the clamp is to hold the material firmly and restrict the plastic

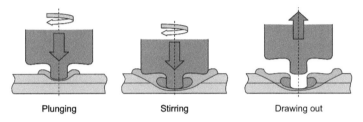

| Plunging | Stirring | Drawing out |

**FIGURE 3.8**
Visual schematic of the three-step friction stir spot welding process (Hovanski, Santella, and Grant 2007). (With kind permission of Elsevier.)

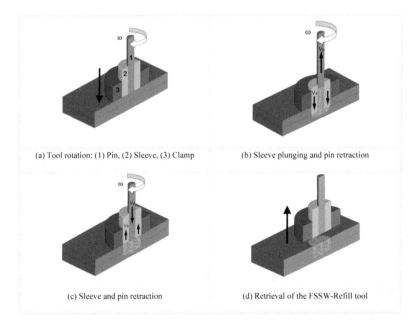

(a) Tool rotation: (1) Pin, (2) Sleeve, (3) Clamp

(b) Sleeve plunging and pin retraction

(c) Sleeve and pin retraction

(d) Retrieval of the FSSW-Refill tool

**FIGURE 3.9**
A schematic illustration of the FSSW-Refill (sleeve plunge variant) showing the four main stages: (a) tool rotation: (1) pin, (2) sleeve, and (3) clamp; (b) sleeve plunging and pin retraction; (c) sleeve and pin retraction; and (d) retrieval of the FSSW-Refill tool (Tier et al. 2013). (With kind permission of Elsevier.)

flow of the material. In contrast, the pin and sleeve rotates and translates accordingly to execute the refill process. Another way to avoid the backlog of conventional FSSW is to have a small amount of traverse movement similar to FSW, which is termed as stitch-FSSW. The objective of this methodology is to produce joints with a larger bonding area for higher strength (Yuan 2008). Swing-FSSW is a new variant of FSSW developed by Hitachi to reduce the size of the exit pinhole (Yuan 2008). The process is similar to conventional FSSW process with an additional step. After plunging step, the tool swings with a larger radius and smaller angular displacement. The swing results in the squeezing of the material located at the end of the welding. Another variant of FSSW is probeless FSSW (P-FSSW), which is more suitable to join thin sheets in comparison to conventional FSSW (Chu et al. 2017; Li et al. 2018). The FSSW review providing an overview of joint macro- and microstructures, property, and process modeling has been reported (Yang, Fu, and Li 2014).

## 3.4 Stationary Shoulder Friction Stir Welding (SSFSW)

To overcome the problem of poor weldability of titanium alloys, which observes throughthickness temperature gradient in conventional FSW process, TWI has developed an innovative welding process named SSFSW. The tooling system of SSFSW is shown in Figure 3.10. In the SSFSW process, a rotating pin located inside a stationary shoulder, which slides over the material surface, carries out the welding process. The contribution by the shoulder toward heat generation and material flow is almost none (Beckman and Sundström 2014). The rotating pin generates frictional heat in the surrounded area. This frictional heat is focused around the pin and is responsible for stirring of the material. The main processing parameters in the SSFSW process include pin rotational speed, traverse speed, plunge depth, and tool geometry (Padhy, Wu, and Gao 2017). As the shoulder is stationary in the SSFSW process, it can also be used for producing fillet welds using solid-state welding

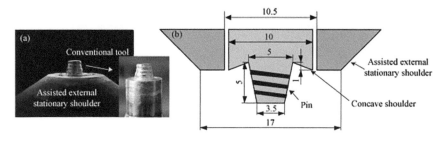

**FIGURE 3.10**
The assisted external stationary shoulder: (a) stationary shoulder system and (b) its dimensions (Ji et al. 2016). (With kind permission of Elsevier.)

process. Various industries have developed their own SSFSW welding methodology. For producing a fillet weld using SSFSW, a separate welding wire is fed directly above the probe which allows creating the fillet.

## 3.5 Bobbin Tool or Self-Reacting Friction Stir Welding (SRFSW)

Self-reacting FSW (SRFSW), also known as Bobbin tool FSW (BTFSW), is a variant of FSW technique, as shown in Figure 3.11a. In SRFSW, the backing plate used in the conventional FSW process is removed due to innovative tool design. It is a full penetration solid-state welding procedure which eliminates the formation of root defects which are quite common in FSW (Singh 2016). The modification in the tool design results in different microstructure than the conventional FSW (Li et al. 2014; Wang et al. 2015, 2016, 2018). The different microstructural zones of the SRFSW process are shown in Figure 3.11b. The different zones are the same as that of the FSW process, which are a nugget, TMAZ, HAZ, and base metal. The only difference over here is that these zones are symmetric across the thickness of the material. The only limitation of the SRFSW process is that it requires a pilot hole for starting the process (Huang et al. 2013). The authors have also tried it for joining aluminum hollow extrusion and have succeeded (Huang et al. 2013). Still, BTFSW is emerging as an important variant of FSW.

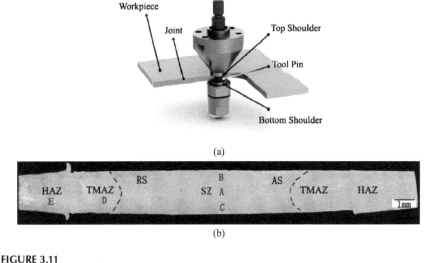

(a)

(b)

**FIGURE 3.11**
(a) Schematic of BT-FSW and (b) a typical macroscopic image of the BT-FSWed joint (Li et al. 2014). (With kind permission of Elsevier.)

## 3.6 Friction Stir Channeling (FSC)

FSW is very well known for its defect-free welds, but in rare cases, a defect called "wormhole" is generated if processing parameters are not selected properly (Nagarajan Balasubramanian 2008). The idea of reproducing this defect in an organized manner and use it for heat exchanger application is termed as FSC. Mishra et al. have shown that a continuous channel can be generated in a single plate by selecting a suitable combination of parameters (Mishra 2005). The main aspects to produce a channel are as follows (Balasubramanian 2008):

  a. the profiled tool is used such that the material flow is upward toward the shoulder;
  b. an initial clearance is provided between the shoulder and the workpiece where the material from the base of the pin is deposited;
  c. the working gap between the shoulder and the workpiece is adjusted in a particular manner to obtain the required size, shape, and integrity of the channel.

The major difference between FSW and FSC is the gap between the shoulder and the base material. In the FSW process, the face of the shoulder touches the surface of the workpiece to generate the forging action required to produce the defect-free weld. During the FSC process, an upward force is generated by a rotating tool with threads (for right-handed threads, clockwise rotation and for left-handed threads, anticlockwise rotation) (Nagarajan Balasubramanian 2008). A cavity is formed by separation of plasticized material around the pin and plasticized material at bottom of the probe. This separated material gets transported into the gap present between the shoulder and the base metal. The shape and size of this cavity can be controlled by adjusting the processing parameters. When a translation motion is provided to the rotating tool, it forms a channel by integrating all the small cavities (Balasubramanian, Mishra, and Krishnamurthy 2009).

## 3.7 Friction Stir Processing (FSP)

Mishra and Mahoney (2001) developed a new material processing technology named FSP based on the basic principle of the FSW process. FSP follows the same working principles of FSW to alter the microstructure and mechanical properties of the base metal rather than joining, as schematically represented in Figure 3.12. One of the significances of FSP process is that it

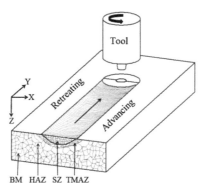

**FIGURE 3.12**
Schematic of friction stir processing (FSP) showing different regions (Yadav and Bauri 2012). (With kind permission of Elsevier.)

can be used for different applications such as grain refinement, superplasticity, surface composite manufacturing, casting and fusion weld repair, compositional homogeneity, and many more. In this section, the discussion is focused on using FSP process for grain refinement, achieving superplasticity, fabricating surface and bulk composites, and casting and fusion weld repair.

### 3.7.1 Grain Refinement

During the FSP process, severe frictional heating and intense plastic deformation of the material in the SZ results in the occurrence of dynamic recrystallization process (Mishra, De, and Kumar 2014). As a result, fine and equiaxed recrystallized grains with uniform size distribution are formed in the SZ (Chang, Du, and Huang 2007). Tool rotation and travel speed during the FSP process mainly govern the heat input, thus resulting grain size in the SZ. The low heat input condition suggests that a combination of low tool rotation and high travel speed restricts the grain coarsening, followed by enhanced grain refinement (Patel, Badheka, and Kumar 2016a,b,d). Tool geometry also plays a significant role in heat generation during FSP (Patel, Badheka, and Kumar 2017a,b). Depending on the cooling rate of the SZ, different grain sizes ranging from the few tens of nanometer to a few hundred nanometers have been achieved (Su, Nelson, and Sterling 2006). The hybrid FSP approach using a different cooling medium such as compressed air, water, and $CO_2$ has demonstrated the further enhancement in the grain refinement of FSPed Al7075 alloy (Patel, Badheka, et al. 2016; Patel et al. 2017). However, there is no study available which directly relates the cooling rate with the grain size distribution for an alloy system. There are various other grain refinement techniques such as equi-channel angular pressing (Valiev and Langdon 2006), torsion under compression (Mishra and Mahoney 2001), multiaxial alternative forging (Noda, Hirohashi, and Funami 2003), and accumulative roll bonding (Tsuji, Shiotsuki, and Saito 1999). When compared with other processes, the FSP is a simple and less time-consuming process, leading to lower processing cost. Second, the dimension of the specimen is rather large and thick compared to another process. Grain refinement of the base metal with a thickness of several tens of millimeters has been achieved via FSP process (Ma 2008).

### 3.7.2 Superplasticity

The ability of a polycrystalline material to exhibit, in a generally isotropic manner, very high elongation prior to failure is termed as superplasticity (Langdon 2009). To achieve structural superplasticity, two things are of prerequisite, namely, grain size (<15 µm) and thermal stability of the fine grain microstructure at high temperature (Ma 2008; Mishra and Ma 2005). Conventionally, thermomechanical processing (TMP) is used to produce fine grain microstructure in commercial aluminum alloys (Mishra and Ma 2005). Conventional TMP process is very complex and time-consuming, which results in increased material cost. Mishra et al. first investigated the super-plastic behavior of Al7075 alloy using FSP technique (Mishra et al. 1999). They observed that FSPed Al 7075 alloy with a grain size of ~3.3 µm exhibited high strain rate superplasticity. A maximum elongation of 1000% was observed at a strain rate of $10^{-2} s^{-1}$ and 490°C. Patel, Badheka, and Kumar (2016c) provide a classic review of FSPed superplasticity in various aluminum alloys.

### 3.7.3 Surface and Bulk Composite Manufacturing

The FSP process produces high strain (approximately 40) and mixing of the material in the SZ (Ma 2008). It is possible to incorporate hard ceramic particles into the metal matrix during FSP process to produce surface/bulk composite materials. Mishra et al. succeeded in fabricating surface composite by incorporating SiC hard particles in a soft aluminum matrix using FSP (Mishra, Ma, and Charit 2003). The SiC powder was mixed with a small amount of methanol and was then applied to the surface of the Al5083 alloy plate. This preplaced SiC layered plate was subjected to FSP. As a result, a plate with a composite layer of thickness between 50 and 200 µm was obtained. The SiC particle showed an excellent bonding with the aluminum matrix. The hardness of the surface composite material with 27 vol% SiC was approximately 173HV which was almost doubled compared to the substrate hardness of 85HV (Mishra, Ma, and Charit 2003). Another way to produce metal-matrix surface/bulk composite via FSP is to produce a groove on the surface in direction of FSP and deposit the material inside it (Ma 2008). The surface composite of Al7075+$B_4C$ has been reported using the FSP process (Rana and Badheka 2017; Rana et al. 2017). Morisada et al. (2006) reported the fabrication of multi-walled carbon nanotubes (MWCNTs)/magnesium alloy (AZ31) composite material by the FSP process. In composite material fabrication, MWCNTs were deposited in a groove (1 mm × 2 mm) on the AZ31 plate.

A fine dispersion of MWCNTs along the thickness of the plate was observed, and a grain with size <500 nm was obtained. The microhardness of the material was improved in comparison to the base material AZ31.

### 3.7.4 Casting and Fusion Weld Repair

Al–Mg–Si–Cu alloys are widely used to cast high-strength components used for automotive and aerospace applications because they offer a good combination of castability and strength (Mishra and Ma 2005). However, some properties are limited in cast alloys such as ductility, toughness, and fatigue resistance. The limitations are because of the porosity, coarse Si particles, and coarse primary aluminum dendrites (Mishra and Ma 2005). Research has been carried out to modify the microstructure of the cast alloys for enhancing the properties such as toughness, ductility, and fatigue resistance using the FSP process. Mishra and Ma (2005) used the FSP technique to modify the microstructure of cast Al–Mg–Si alloy. Experimental observation showed that microstructural modifications of cast Al–Mg–Si alloy improved mechanical properties, in particular, fatigue and ductility. There are also some developments to repair defects like porosity, blowhole, and crack produced by fusion welding process using FSP techniques (Prasad Rao, Janaki Ram, and Stucker 2010; Borrego et al. 2014; Costa et al. 2014). Observations showed that microstructure of the material was highly modified using FSP process and mechanical properties enhanced due to the defect-free FSPed weld zones.

## 3.8 Summary

It can be summarized from the above survey that a strong potential exists for the important FSW variants that can gain substantial focus in the coming time. Some of the important variants include the assisted or hybrid FSW process where the external source of energy is involved in the processes such as PFSW, GFSW, LFSW, GTAFSW, EAFSW, I-FSW, and UAFSW; processes such as BTFSW and SSFSW where the modified tooling system is incorporated; and lastly the processes such as FSSW, FSC, and FSP where the slight modification is carried out.

## Acknowledgments

The authors thank the National Natural Science Foundation of China [grant number 51574196]; National Key Research and Development Program of China [grant number 2016YFB1100104]; the State Key Laboratory of Solidification Processing [grant number 122-QZ-2015]; and the 111 Project [grant number B08040] for their financial support.

---

## References

AWS. 2007. *Welding Handbook*, vol. 3. Miami, FL: American Welding Society. doi:10.1017/CBO9781107415324.004.

Balasubramanian, N., 2008. Friction stir channeling: An innovative technique for heat exchanger manufacturing. *Doctoral Dissertations*, Missouri University of Science and Technology.

Balasubramanian, N., R. S. Mishra, and K. Krishnamurthy. 2009. "Friction stir channeling: Characterization of the channels." *Journal of Materials Processing Technology* no. 209 (8):3696–3704. doi:10.1016/j.jmatprotec.2008.08.036.

Bang, H., H. Bang, G. Jeon, I. Oh, and C. Ro. 2012. "Gas tungsten arc welding assisted hybrid friction stir welding of dissimilar materials Al6061-T6 aluminum alloy and STS304 stainless steel." *Materials & Design* no. 37:48–55.

Beckman, A., and M. Sundström. 2014. Development of stationary shoulder for friction stir welding. Chalmers University of Technology.

Borrego, L. P., J. D. Costa, J. S. Jesus, A. R. Loureiro, and J. M. Ferreira. 2014. "Fatigue life improvement by friction stir processing of 5083 aluminium alloy MIG butt welds." *Theoretical and Applied Fracture Mechanics* no. 70:68–74.

Chang, C. I., X. H. Du, and J. C. Huang. 2007. "Achieving ultrafine grain size in Mg-Al-Zn alloy by friction stir processing." *Scripta Materialia* no. 57 (3): 209–212. doi:10.1016/j.scriptamat.2007.04.007.

Chang, W.-S., S. R. Rajesh, C.-K. Chun, and H.-J. Kim. 2011. "Microstructure and mechanical properties of hybrid laser-friction stir welding between AA6061-T6 Al alloy and AZ31 Mg alloy." *Journal of Materials Science & Technology* no. 27 (3):199–204.

Choi, D. H., C. Y. Lee, B. W. Ahn, J. H. Choi, Y. M. Yeon, K. Song, S. G. Hong, W. B. Lee, K. B. Kang, and S. B. Jung. 2011. "Hybrid friction stir welding of high-carbon steel." *Journal of Materials Science and Technology* no. 27 (2): 127–130. doi:10.1016/S1005-0302(11)60037-6.

Chu, Q., W. Y. Li, X. W. Yang, J. J. Shen, Y. B. Li, and W. B. Wang. 2017. "Study of process/structure/property relationships in probeless friction stir spot welded AA2198 Al-Li alloy." *Welding in the World* no. 61 (2):291–298.

Costa, J. D. M., J. S. Jesus, A. Loureiro, J. A. M. Ferreira, and L. P. Borrego. 2014. "Fatigue life improvement of mig welded aluminium T-joints by friction stir processing." *International Journal of Fatigue* no. 61:244–254.

Ferrando, W A. 2008. The Concept of Electrically Assisted Friction Stir Welding (EAFSW) and application to the processing of various metals." *Naval Surface Warfare Center Carderock Division Bethesda, Survivability, Structure, and Materials Department.* http://www.dtic.mil/cgi-bin/GetTRDoc?AD=ADA319658APPROVED.

Gallais, C., A. Simar, D. Fabregue, A. Denquin, G. Lapasset, B. de Meester, Y. Brechet, and T. Pardoen. 2007. "Multiscale analysis of the strength and ductility of AA 6056 aluminum friction stir welds." *Metallurgical and Materials Transactions A* no. 38 (5):964–981.

Gibson, B. T., D. H. Lammlein, T. J. Prater, W. R. Longhurst, C. D. Cox, M. C. Ballun, K. J. Dharmaraj, G. E. Cook, and A. M. Strauss. 2014. "Friction stir welding: Process, automation, and control." *Journal of Manufacturing Processes* no. 16 (1):56–73.

Hovanski, Y., M. L. Santella, and G. J. Grant. 2007. "Friction stir spot welding of hot-stamped boron steel." *Scripta Materialia* no. 57 (9):873–876.

Huang, Y. X., L. Wan, S. X. Lv, and J. C. Feng. 2013. "Novel design of tool for joining hollow extrusion by friction stir welding." *Science and Technology of Welding and Joining* no. 18 (3): 239–246. doi:10.1179/1362171812Y.0000000096.

Ji, S., Z. Li, L. Zhang, Z. Zhou, and P. Chai. 2016. "Effect of lap configuration on magnesium to aluminum friction stir lap welding assisted by external stationary shoulder." *Materials & Design* no. 103:160–170.

Langdon, T.G. 2009. "Seventy-five years of superplasticity: Historic developments and new opportunities." *Journal of Materials Science* no. 44 (22): 5998–6010. doi:10.1007/s10853-009-3780-5.

Li, W. Y., Q. Chu, X. W. Yang, J. J. Shen, A. Vairis, and W. B. Wang. 2018. "Microstructure and morphology evolution of probeless friction stir spot welded joints of aluminum alloy." *Journal of Materials Processing Technology* no. 252:69–80.

Li, W. Y., T. Fu, L. Hütsch, J. Hilgert, F. F. Wang, J. F. dos Santos, and N. Huber. 2014. "Effects of tool rotational and welding speed on microstructure and mechanical properties of bobbin-tool friction-stir welded Mg AZ31." *Materials & Design* no. 64:714–720.

Luo, J., W. Chen, and G. Fu. 2014. "Hybrid-heat effects on electrical-current aided friction stir welding of steel, and Al and Mg alloys." *Journal of Materials Processing Technology* no. 214 (12):3002–3012.

Ma, Z. Y. 2008. "Friction stir processing technology: A Review." *Metallurgical and Materials Transactions A* no. 39A: 642–658. doi:10.1007/s11661-007-9459-0.

Mishra, R. S. 2005. Integral channels in metal components. US 6923362.

Mishra, R. S., and M. W. Mahoney. 2001. "Friction stir processing: A new grain refinement technique to achieve high strain rate superplasticity in commercial alloys." *Materials Science Forum* no. 357:507–514.

Mishra, R. S., and Z. Y. Ma. 2005. "Friction stir welding and processing." *Materials Science and Engineering R* no. 50 (1–2): 1–78. doi:10.1016/j.mser.2005.07.001.

Mishra, R. S., Z. Y. Ma, and I. Charit. 2003. "Friction stir processing: A novel technique for fabrication of surface composite." *Materials Science and Engineering A* no. 341 (1–2): 307–310. doi:10.1016/S0921-5093(02)00199-5.

Mishra, R. S., P. S. De, and N. Kumar. 2014. "Friction stir processing." In *Friction Stir Welding and Processing: Science and Engineering*, 259–296. Cham: Springer International Publishing.

Mishra, R. S., M. W. Mahoney, S. X. McFadden, N. A. Mara, and A. K. Mukherjee. 1999. "High strain rate superplasticity in a friction stir processed 7075 Al alloy." *Scripta Materialia* no. 42 (2): 163–168. doi:10.1016/S1359-6462(99)00329-2.

Morisada, Y., H. Fujii, T. Nagaoka, and M. Fukusumi. 2006. "MWCNTs/AZ31 surface composites fabricated by friction stir processing." *Materials Science and Engineering A* no. 419 (1–2): 344–348. doi:10.1016/j.msea.2006.01.016.

Nandan, R, T. Debroy, and H. K. D. H. Bhadeshia. 2008. "Recent advances in friction stir welding: Process, weldment structure and properties." *Progress in Materials Science* no. 53 (6): 980–1023. doi:10.1016/j.pmatsci.2008.05.001.

Noda, M., M. Hirohashi, and K. Funami. 2003. "Low temperature superplasticity and its deformation mechanism in grain refinement of Al-Mg alloy by multi-axial alternative forging." *Material Transactions* no. 44 (11): 2288–2297. doi:10.2320/jinstmet1952.67.2_98.

Padhy, G. K., C. S. Wu, and S. Gao. 2015. "Auxiliary energy assisted friction stir welding: Status review." *Science and Technology of Welding and Joining* no. 20 (8): 631–649. doi:10.1179/1362171815Y.0000000048.

Patel, V. V., V. J. Badheka, and A. Kumar. 2016a. "Influence of friction stir processed parameters on superplasticity of Al-Zn-Mg-Cu alloy." *Materials and Manufacturing Processes* no. 31 (12):1573–1582.

Patel, V. V., V. J. Badheka, and A. Kumar. 2016b. "Cavitation in friction stir processing of Al-Zn-Mg-Cu alloy." *International Journal of Mechanical Engineering and Robotics Research* no. 5 (4):317–321. doi: 10.18178/ijmerr.5.4.317-321.

Patel, V. V., V. J. Badheka, and A. Kumar. 2016c. "Friction stir processing as a novel technique to achieve superplasticity in aluminum alloys: Process variables, variants, and applications." *Metallography, Microstructure, and Analysis* no. 5 (4):278–293.

Patel, V. V., V. J. Badheka, and A. Kumar. 2016d. "Effect of velocity index on grain size of friction stir processed Al-Zn-Mg-Cu alloy." *Procedia Technology* no. 23:537–542. doi: 10.1016/j.protcy.2016.03.060.

Patel, V. V., V. J. Badheka, and A. Kumar. 2017a. "Influence of pin profile on the tool plunge stage in friction stir processing of Al–Zn–Mg–Cu alloy." *Transactions of the Indian Institute of Metals* no. 70 (4):1151–1158. doi: 10.1007/s12666-016-0903-y.

Patel, V. V., V. J. Badheka, and A. Kumar. 2017b. "Effect of polygonal pin profiles on friction stir processed superplasticity of AA7075 alloy." *Journal of Materials Processing Technology* no. 240:68–76. doi: 10.1016/j.jmatprotec.2016.09.009.

Patel, V. V., V. J. Badheka, S. R. Zala, S. R. Patel, U. D. Patel, and S. N. Patel. 2016. Effects of various cooling techniques on grain refinement of aluminum 7075-T651 during friction stir processing. *Paper Read at ASME 2016 International Mechanical Engineering Congress and Exposition*, Phoenix, AZ: American Society of Mechanical Engineers.

Patel, V. V., D. J. Sejani, N. J. Patel, J. J. Vora, B. J. Gadhvi, N. R. Padodara, and C. D. Vamja. 2016. "Effect of tool rotation speed on friction stir spot welded AA5052-H32 and AA6082-T6 dissimilar aluminum alloys." *Metallography, Microstructure, and Analysis* no. 5 (2):142–148.

Patel, V. V., V. J. Badheka, U. Patel, S. Patel, S. Patel, S. Zala, and K. Badheka. 2017. "Experimental investigation on hybrid friction stir processing using compressed air in aluminum 7075 alloy." *Materials Today: Proceedings* no. 4 (9):10025–10029. doi: 10.1016/j.matpr.2017.06.314.

Prasad Rao, K., G. D. Janaki Ram, and B. E. Stucker. 2010. "Effect of friction stir processing on corrosion resistance of aluminum–copper alloy gas tungsten arc welds." *Materials & Design* no. 31 (3):1576–1580.

Rana, H., and V. J. Badheka. 2017. "Elucidation of the role of rotation speed and stirring direction on AA 7075-$B_4C$ surface composites formulated by friction stir processing." *Proceedings of the Institution of Mechanical Engineers, Part L: Journal of Materials: Design and Applications.* doi: 10.1177/1464420717736548.

Rana, H., V. J. Badheka, A. Kumar, and A. Satyaprasad. 2017. "Strategical parametric investigation on manufacturing of Al–Mg–Zn–Cu alloy surface composites using FSP." *Materials and Manufacturing Processes* no. 33:1–12.

Shi, L., C. S. Wu, and X. C. Liu. 2015. "Modeling the effects of ultrasonic vibration on friction stir welding." *Journal of Materials Processing Technology* no. 222:91–102.

Singh, P. 2016. "A three-dimensional fully coupled thermo-mechanical model for self-reacting friction stir welding of aluminium AA6061 sheets." *Journal of Physics* doi:10.1088/1742-6596/759/1/012047.

Su, J. Q., T. W. Nelson, and C. J. Sterling. 2006. "Grain refinement of aluminum alloys by friction stir processing." *Philosophical Magazine* no. 86 (1): 1–24. doi:10.1080/14786430500267745.

Thomas, W. M., E. D. Nicholas, J. C. Needham, M. G. Murch, P. Templesmith, and C. J. Dawes. December 1991. Great Britain Patent Application No. 9125978.8.

Tier, M. D., T. S. Rosendo, J. F. dos Santos, N. Huber, J. A. Mazzaferro, C. P. Mazzaferro, and T. R. Strohaecker. 2013. "The influence of refill FSSW parameters on the microstructure and shear strength of 5042 aluminium welds." *Journal of Materials Processing Technology* no. 213 (6):997–1005.

Tsuji, N., K. Shiotsuki, and Y. Saito. 1999. "Superplasticity of ultra-fine grained Al-Mg alloy by accumulative roll-bonding." *Materials Transactions* no. 40:765–771.

Valiev, R. Z., and T. G. Langdon. 2006. "Principles of equal-channel angular pressing as a processing tool for grain refinement." *Progress in Materials Science* no. 51 (7): 881–981. doi:10.1016/j.pmatsci.2006.02.003.

Vijendra, B., and A. Sharma. 2015. "Induction heated tool assisted friction-stir welding (i-FSW): A novel hybrid process for joining of thermoplastics." *Journal of Manufacturing Processes* no. 20:234–244.

Wang, F. F., W. Y. Li, J. Shen, S. Y. Hu, and J. F. dos Santos. 2015. "Effect of tool rotational speed on the microstructure and mechanical properties of bobbin tool friction stir welding of Al–Li alloy." *Materials & Design* no. 86:933–940.

Wang, F. F., W. Y. Li, J. Shen, Q. Wen, and J. F. dos Santos. 2018. "Improving weld formability by a novel dual-rotation bobbin tool friction stir welding." *Journal of Materials Science & Technology* no. 34:135–139.

Wang, F. F., W. Y. Li, J. Shen, Z. H. Zhang, J. L. Li, and J. F. dos Santos. 2016. "Global and local mechanical properties and microstructure of Bobbin tool friction-stir-welded Al–Li alloy." *Science and Technology of Welding and Joining* no. 21 (6):479–483.

Yadav, D., and R. Bauri. 2012. "Effect of friction stir processing on microstructure and mechanical properties of aluminium." *Materials Science and Engineering: A* no. 539:85–92.

Yaduwanshi, D. K., S. Bag, and S. Pal. 2016. "Numerical modeling and experimental investigation on plasma-assisted hybrid friction stir welding of dissimilar materials." *Materials & Design* no. 92:166–183.

Yang, X. W., T. Fu, and W. Y. Li. 2014. "Friction stir spot welding: A review on joint macro-and microstructure, property, and process modelling." *Advances in Materials Science and Engineering* no. 2014:1–12.

Yuan, W. 2008. "Friction stir spot welding of aluminum alloys." *Materials Testing* doi:10.3139/120.110752.

# 4

# *Different Methodologies for the Parametric Optimization of Welding Processes*

**Jaykumar J. Vora**

*Pandit Deendayal Petroleum University*

**Kumar Abhishek and PL. Ramkumar**

*Institute of Infrastructure Technology Research and Management*

## CONTENTS

4.1 Introduction ........................................................................................55
4.2 Optimization Methodology ............................................................57
    4.2.1 Optimization Process .............................................................57
    4.2.2 Optimization Methods ...........................................................57
    4.2.3 Taguchi's Optimization .........................................................57
    4.2.4 Response Surface Methodology ...........................................66
    4.2.5 Meta-Heuristics Methods ......................................................66
4.3 Summary ............................................................................................70
4.4 Future Scope ......................................................................................71
References .....................................................................................................71

## 4.1 Introduction

The manufacturing sector is changing and expanding by leaps and bound in the ever-changing world. The processes which were considered novel before a decade have now been transformed into entirely new variants. Manufacturing, as a whole, is a very vast domain comprising of several processes such as turning, milling, welding, casting forming, etc. Out of those, welding is one of the processes which is used extensively in almost every sector including automotive, oil and gas industry, aerospace industry, etc. Welding is given a special consideration by almost all the nations across the globe, as the listed sectors are the ones that guide any nation to the path of progress and prosperity. Hence, these processes, in general, require an increased control and a reliable outcome in terms of quality of the welded part. In general, the welded joint quality depends on the specific input parameters during welding. Given that, almost all welding processes such as arc welding and solid-state welding processes are very complex and dependent on a lot of variables. Also, the output quality parameters measured and expected from these processes are also complex and interdependent. Thus, these processes can be termed as a multi-input multi-output process. Unfortunately, a mutual challenge faced by the manufacturers in the domain is the devising the exact input process parameters setting to get required weld quality and an accurate control of these parameters, to obtain a sound welded joint with the required bead morphology and strength at the same time, with minimum undesirable defects, residual stresses, and distortion. Conventionally, for a manufacturer, it is essential to decide the input process parameters for every new components/joints to be welded with the required specifications. In addition to that, in the modern era, the selection of the base materials is taking a paradigm shift. Everyday a new material or a combination of more than one material, which performs better in some aspect as compared to the existing material, is developed. This makes the task of the manufacturer even tougher as these materials with diverse properties and unknown response to the welding have to be successfully joined to manufacture components. The scenario becomes even more challenging as they are bound by constraints of time, cost, and material usage. The general approach of the manufacturer is to select a working window for the input parameters based on the prior knowledge and skill of the engineer/welding operator and then deposited the welds by trial and error approach using a specific set of input parameters. This would be followed by the testing of welds to assess its conformity of specification. However, this approach is always time-consuming, unreliable, and costly as a lot of experiments might be needed to reach the required quality.

In order to encounter this, the researchers and manufacturers in the domain opt for the statistical and numerical approaches to solve the problem. This technique, in a nutshell, dictates the experimentation methodology to reach an optimum level of input parameters for required quality specification. The number of trials to be taken as per these techniques is more reliable and less as compared to the previous approach. Many of these techniques establish a mathematical relationship between the input and output parameters which can be used to understand the effect and role of specific input parameter on a given quality trait. In the past few years, a different design of experiment (DoE) techniques has been used to carry out such optimization. The use of evolutionary algorithms and computational network has also grown rapidly which have been also adapted for many applications in different areas. Thus, an attempt has been made through this chapter to summarize and review the different methodologies followed to optimize the input parameters of different welding processes. Several techniques such as the Taguchi method, response surface methodology, artificial neural network (ANN), and genetic algorithm (GA) individually or as a combination have been experimented with different welding processes on different steels which have been included herewith.

## 4.2 Optimization Methodology

### 4.2.1 Optimization Process

The welding processes involve a large number of process parameters; however, based on the engineer's basic knowledge, available machine capabilities, and desired quality specification required, certain factors are kept constant so as to give desired effect on output. Several instances have been reported where 3 to 6 most influential parameters are selected from all the possible parameters and further optimized. It is imperative to note that a tentative working window based on the machine availability is also selected for further steps. The next step is an experimentation matrix which is formulated based on the selected optimization technique, followed by taking actual welding trials as per this matrix. The next step is to measure and analyze the required quality output specification required. After the measurement and record of these data, an appropriate analysis which is specific to the optimization method which is selected is carried out and the final outcome is the set of optimized parameters for obtaining the maximum values of the output quality parameter selected earlier. Confirmatory trials are also carried out to check the robustness of the technique.

### 4.2.2 Optimization Methods

In its most basic terms, optimization can be defined as a methodical process using design constraints and bounds which allows the user to locate the optimum solution. Lockhart and Johnson (2000) presented the optimization as a methodical approach, using design constraints and bounds to find the most effective/favorable value/condition. Over the past few decades, major efforts on optimization of manufacturing processes (primarily welding) have been made using the techniques as shown in Figure 4.1. However, the majority of the studies have reported the combination of one or more techniques to optimize the parameters. A comprehensive survey of these approaches has been presented in a tabulated form in Table 4.1.

### 4.2.3 Taguchi's Optimization

Taguchi's method or Taguchi's orthogonal array (OA) is popularly known as a systematic application of DoE concept wherein maximum information is extracted from minimum experimental effort conducted as per developed designs and concepts. An experimental design is a technique of laying out a plan of experiments prior to executing the trials after careful review of the objective of experiments and the affecting input process parameters.

Taguchi method uses a special design of OAs to study the entire process parameter space with a reduced number of experiments. Orthogonality

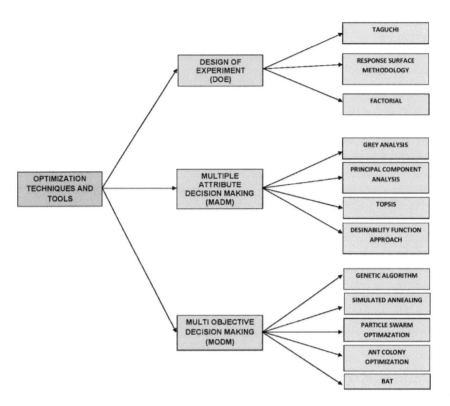

**FIGURE 4.1**
General optimization techniques.

means that every selected factor is independently evaluated and the effect of one factor does not interfere with the estimation of the influence of another factor (Wang and Northwood 2008). A set of experimental trials are devised by this technique and further those experiments are conducted and responses are measured and recorded. After conducting the experiments, the optimum test parameter configuration from the design matrix was determined. To further analyze the results, signal-to-noise (S/N) ratio which is a statistical tool for the performance measurement is calculated. The S/N equation depends on the criterion for the quality characteristic to be optimized, i.e., higher the better, lower the better, and normal the better. The loss function is also converted in the S/N ratio, and for each level of the process parameters, S/N ratio is calculated based on S/N analysis. Larger S/N ratio corresponds to the better quality characteristic and corresponding levels of parameters are termed as optimized parameters. Narwadkar and Bhosle (2016), for example, optimized the gas metal arc welding (GMAW) process parameters for controlling the angular distortion in the resultant welded joints. In the study, three factors, namely,

**TABLE 4.1**

Research Work Carried Out in the Optimization of the Welding Process

| Year/Author | Welding Method | Base Metal | Input Parameters | Output Response | Optimization Technique Used |
|---|---|---|---|---|---|
| Huang (2010) | Gas metal arc welding | AISI 1020 carbon steel | Arc voltage, welding current, welding speed, Joint gap | Weld-bead geometry, weld area, weld penetration | Taguchi method |
| Magudeeswaran et al. (2014) | Activated tungsten inert gas welding | Duplex stainless steel (ASTM/UNS: S32205) | Electrode gap, travel speed, welding current, welding voltage | Width of the weld bead, weld penetration, depth-to-width ratio | Taguchi method |
| Pujari Srinivasa Rao, Rao, and Priyangeli (2016) | Activated tungsten inert gas welding | Titanium—Ti-6Al-4V | Flux coating, welding current, welding speed | Depth penetration, heat-affected zone | Taguchi method |
| Kumar and Sundarrajan (2009) | Pulsed tungsten inert gas welding | AA 5456 aluminum alloy | Peak current, base current, welding speed, pulse frequency | Yield strength, percent elongation, hardness, ultimate tensile strength | Taguchi method with regression analysis |
| Narwadkar and Bhosle (2016) | Gas metal arc welding | Fe410WA steel | Welding current, welding voltage, gas flow rate | Angular distortion | Taguchi method |
| Kangazian and Shamanian (2016) | Pulsed tungsten inert gas welding | AISI 304 stainless steel and St 52 carbon steel plates | Pulse current, background current, pulse frequency | Micro-hardness, corrosion property | Taguchi method |
| Juang and Tarng (2002) | Tungsten inert gas welding | Stainless steel | Welding arc gap, gas flow rate, welding current, welding speed | Weld pool geometry | Taguchi method |
| Bhattacharya and Singla (2017) | Tungsten inert gas welding | AISI 316 and AISI 4340 | Welding current, Shielding gas, gas flow rate, filler material, pre-heating, post-heating | Tensile strength, toughness, micro-hardness | Taguchi method with factorial design |
| Lin and Yan (2014) | Activated gas metal arc welding | Aluminum alloy | Welding voltage, travel speed, flow rate of gas, mixed component fluxes | Weld penetration, depth-to-width ratio, weld area | Taguchi with grey relational analysis |

*(Continued)*

**TABLE 4.1 (*Continued*)**

Research Work Carried Out in the Optimization of the Welding Process

| Year/Author | Welding Method | Base Metal | Input Parameters | Output Response | Optimization Technique Used |
|---|---|---|---|---|---|
| Asif, Shrikrishna, and Sathiya (2016) | Friction welding | Duplex stainless steel | Heating pressure, heating time, upsetting pressure, upsetting time | Corrosion rate, tensile strength, impact toughness, hardness | Taguchi with grey relational analysis |
| Lin (2013) | Activated tungsten inert gas welding | Inconel 718 alloy | Arc length, welding speed, flow rate of argon gas, welding current, angle of electrode tip, mixed flux type | Depth-to-width ratio, weld-bead penetration depth | Taguchi with grey relational analysis |
| Sahu and Pal (2015) | Friction stir welding | AM20 magnesium alloy | Plunge depth, tool rotational, speed welding speed shoulder diameter | Ultimate tensile strength, yield strength, percent elongation, compressive stress | Taguchi with grey relational analysis |
| Sathiya et al. (2011) | Laser welding | AISI 904 L super austenitic stainless steel | Beam power, travel speed, focal position | Bead width, weld penetration | Taguchi with grey relational analysis |
| Sathiya, Jaleel, and Katherasan (2010) | Laser welding | Super austenitic stainless steel | Beam power, travel speed, focal position | Bead width, weld penetration | Taguchi with grey relational analysis |
| Hsiao, Tarng, and Huang (2007) | Plasma arc welding | SUS316 stainless steel plate | Torch stand-off, welding current, welding speed, plasma gas flow rate | Weld penetration, welding groove width, front-side undercut | Taguchi with grey relational analysis |
| Abhulimen and Achebo (2014) | Tungsten inert gas welding | Mild steel pipe | Welding voltage, gas flow rate, welding current, electrode diameter | Ultimate tensile strength, yield strength, strain | Artificial neural network |
| Anand et al. (2015) | Friction welding | Incoloy 800H | Heating pressure, heating time, upsetting pressure, upsetting time | Tensile strength, hardness, burn off length | Artificial neural network |

*(Continued)*

**TABLE 4.1 (*Continued*)**

Research Work Carried Out in the Optimization of the Welding Process

| Year/Author | Welding Method | Base Metal | Input Parameters | Output Response | Optimization Technique Used |
|---|---|---|---|---|---|
| Xu et al. (2015) | Gas metal arc welding | Low carbon steel | Wire feed rate, travel speed, dwell time, oscillating amplitude, welding position | Sidewall penetration, weld penetration, convex degree, weld height | RSM-based Central composite design |
| Zhao et al. (2014) | Resistance spot welding | Titanium alloy | Welding time, welding current, electrode force | Nugget diameter, penetration rate, shear strength, failure energy | RSM-based Box–Behnken design with principal component analysis |
| Santana et al. (2017) | Friction stir spot welding | Al–Mg–Si aluminum alloy | Rotational speed, plunge depth, retracting rate | Rotational speed, plunge depth, retracting rate | RSM-based Box–Behnken design |
| Vasantharaja and Vasudevan (2018) | Activated tungsten inert gas welding | RAFM steel | Welding current, torch speed, arc gap, tip angle | Weld penetration, bead width, heat-affected zone | RSM-based Central composite design |
| Palani and Murugan (2007) | Flux-cored arc welding | Stainless steel | Welding current, welding speed, nozzle-to-plate distance | Percent dilution, weld penetration, coefficient of internal shape of welds, coefficient of external shape of welds | RSM-based Central composite design |
| Kiaee and Aghaie-Khafri (2014) | Tungsten inert gas welding | A516-70 carbon steel plates | Welding current, welding speed, gas flow rate | Weld hardness, HAZ hardness, tensile strength | RSM |
| Boulahem, Salem, and Bessrour (2015) | Friction stir welding | AA2017 aluminum alloy | Rotation speed, traverse speed, tool shoulder, diameter | Surface roughness | Taguchi and RSM |
| Muhammad et al. (2013) | Resistance spot welding | Low carbon steel | Welding current, weld time cycle, hold time | Radius of weld nugget, width of HAZ | Taguchi and RSM |

*(Continued)*

**TABLE 4.1 (*Continued*)**

Research Work Carried Out in the Optimization of the Welding Process

| Year/Author | Welding Method | Base Metal | Input Parameters | Output Response | Optimization Technique Used |
|---|---|---|---|---|---|
| Lin and Chou (2010) | Tungsten inert gas welding | Not specified | Electrode angle, welding current, travel speed, proportion of mixed flux | Weld-bead geometry | Taguchi method and neural-genetic approach |
| Rao and Kalyankar (2013) | Submerged arc welding | Cr–Mo–V steel | Welding current, welding voltage, welding speed, wire feed | Weld-bead width, Weld reinforcement, weld penetration, tensile strength, weld hardness | Taguchi with teaching–learning-based optimization algorithm |
| Subashini, Madhumitha, and Vasudevan (2012) | Activated tungsten inert gas welding | 9Cr-1Mo Steel | Welding current, torch speed, welding voltage | Weld penetration, heat-affected zone | Genetic algorithm |
| Nagaraju et al. (2016) | Activated tungsten inert gas welding | 9Cr-1Mo Steel | Welding current, travel speed, tip angle, arc gap | Weld penetration, heat-affected zone | RSM-based central composite design and genetic algorithm |
| Fuzeau, Vasudevan, and Maduraimuthu (2016) | Activated Tungsten inert gas welding | RAFM steel | Welding current, torch speed, welding voltage | Weld penetration, heat-affected zone | Genetic algorithm with artificial neural network model |
| Correia et al. (2005) | Gas metal arc welding | Mild steel | Reference voltage, wire feed speed welding speed | Weld penetration, deposition efficiency, bead width, bead reinforcement | Genetic algorithms and response surface methodology |
| Sathiya, Ajith, and Soundararajan (2013) | Gas metal arc welding | AISI 904 L stainless steel | Gas flow rate, welding voltage, travel speed, wire feed rate | Bead height, bead width, weld penetration | RSM-based Box–Behnken design and genetic algorithm |

*(Continued)*

**TABLE 4.1 (*Continued*)**

Research Work Carried Out in the Optimization of the Welding Process

| Year/Author | Welding Method | Base Metal | Input Parameters | Output Response | Optimization Technique Used |
|---|---|---|---|---|---|
| Dhas and Kumanan (2011) | Submerged arc welding | Mild steel plates | Welding current, arc voltage, welding speed, electrode stick-out | Weld-bead width | Genetic algorithm and particle swarm optimization algorithm |
| Katherasan et al. (2013) | Flux-cored arc welding | AISI 316L(N) stainless steel | Arc wire feed rate, welding voltage, travel speed, torch angle | Weld penetration, bead width, bead reinforcement | Genetic algorithm and simulated annealing algorithm |
| Deepandurai and Parameshwaran (2016) | Friction stir welding | Cast AA7075/SicP composite | Spindle speed, traveling speed, downward force, percentage of SiC | Ultimate tensile strength, percentage elongation | RSM-based central composite design and fuzzy grey relational analysis (FGRA) |
| Udayakumar et al. (2014) | Friction welding | Duplex stainless steel (UNS S32760) | Friction force, upset force, burn of length | Corrosion current, impact strength | Genetic algorithm with the RSM model and Pareto front |

welding current, welding voltage, and shielding gas flow rate in the range of 180–280 A, 25–31 V, and 15–20 LPM, respectively, were initially selected to be optimized. The authors selected the three levels of the selected input parameters and carried out the experimentation according to the L9 OA. It is worth mentioning that in the absence of the Taguchi method in order to reach the optimized level of the parameters which can yield minimum distortion, an indefinite number of the experimental trials have to be taken as per trial-and-error method. This approach will be costly and time-consuming as well as unreliable as it is a possibility that better results may be achieved at another parameter setting not experimented. Even from Table 4.1, if each and every combination has to experiment, it would take $3^3$, i.e., 27 trials. Furthermore, the Taguchi optimization method successfully optimized the welding parameters and predicted the distortion of 0.108° at the following parameters: current—180 A; voltage—25 V; and gas flow rate—20 LPM. In the confirmatory test taken at the suggested parameters, angular distortion of 0.112° was obtained which was 96.43% close to predict. Magudeeswaran et al. (2014) also carried out the Taguchi optimization by selecting four input parameters, namely, welding current, welding voltage, electrode gap and travel speed to optimize bead width, and depth of penetration and aspect ratio for activated TIG (A-TIG) welding process. The authors selected the L9 OA and further carried out the ANOVA and pooled ANOVA to optimize the parameter as well as to analyze the effect of the input parameters on output responses. The predicted and confirmatory result values were in close agreement with each other. It was also analyzed that the electrode gap was the most dominant factor. A similar study was conducted by Kangazian and Shamanian (2016) on the optimization of pulsed TIG welding process parameters on the hardness and corrosion potential of the resultant weld joints. Three factors (pulse current, background current, and pulse frequency) at two levels and the L4 OA were selected. ANOVA and pooled ANOVA were also carried out, and a close agreement was found between the optimized predicted parameters and confirmatory experiments. Thus, the Taguchi method has been used and implemented fairly well with the welding problems. However, one of the disadvantages of this technique is that for each of the output response parameters to be optimized, the Taguchi method might give a different set of optimized parameters as obtained by Pujari Srinivasa Rao, Rao, and Priyangeli (2016) in their study. The authors used the Taguchi method for optimization of the weld penetration as well as the heat-affected zone (HAZ) by designing an L9 OA. However, they achieved a different set of parameter setting for both the output responses. This situation falls under the category of single-objective optimization but creates a predicament for the end user as to which parameter setting to be used. The Taguchi method can be a good solution for small and target problems; however, for the application which requires the simultaneous optimization of the response parameter, an additional methodology needs to be incorporated.

In order to encounter the same, the Taguchi method is often used with the grey relational analysis (GRA) technique wherein experimental data are first normalized in the range of 0–1. Based on the normalized data, the grey relational coefficient is calculated to represent the correlation between desired and actual experimental data. This is followed by generation of overall grey relational grade by averaging the grey relational coefficient for the respective responses. Thus, the multi-response problem can be converted into a single-response process optimization problem with overall grey relational grade (GRG) as the objective function. Then the last step is to perform ANOVA analysis and prediction of optimal GRG (Sahu and Pal 2015). According to Hakan et al. (2010), the Taguchi method was found to be a very effective tool for process optimization under a limited number of experimental runs. The study showed that the tensile strength and elongation of the welded aluminum alloy were improved by employing the Taguchi method with GRA. Vijayan et al. (2010) carried out the optimization of process parameters in FSW of aluminum alloy with multiple responses based on OA with GRA. Their objective was to investigate the optimum levels of the process parameters which imparts maximum tensile strength and consumes minimum power. In a study by Kasman (2013), multi-response optimization for dissimilar FSW of aluminum alloys was carried out and established that Taguchi method with GRA technique successfully optimized the process parameters. In an independent study, the Taguchi method with GRA was successfully implemented by Hsiao, Tarng, and Huang (2007) to optimize the multiple-response characteristics of plasma arc welding on a 4-mm-thick SS plate. Three responses such as undercut, root penetration, and the welding groove width were measured to identify the optimal welding parameter combination for improvement of welding quality. An improvement in the optimized results was also reported. The GRA technique in conjugation with the Taguchi method was used effectively owing to the ease with which the analysis is carried out. However, further, regression analysis was implemented in combination with the Taguchi method in order to develop mathematical models depicting the input–output relationship of the selected parameters in order to have a better understanding of the process. Several authors have performed theses techniques to develop mathematical models and empirical relationship with between different input process parameters and weld-bead morphological features (Yang, Bibby, and Chandel 1993, Yang, Chandel, and Bibby 1993). Kumar and Sundarrajan (2009) optimized the pulsed gas tungsten arc welding (GTAW) process parameters for welding of aluminum alloy. They used the Taguchi method along with regression analysis to develop the equations for the selected responses. Similar results were obtained by Sathiya et al. (2011) while optimizing the process parameters for the laser beam welding parameters. Despite the fact that regression analysis established the input–output relationship, the generated models were adequate for selecting the window of the input parameters as well as for depicting the linear characteristics (Dhas and Dhas 2012). The complexity of the welding process involves a lot of nonlinear characteristics, and hence, other methods are needed to be incorporated such as response surface methodology (RSM).

### 4.2.4 Response Surface Methodology

RSM is a collection of mathematical and statistical techniques that are useful for the modeling and analysis of problems in which output or response influenced by several input variables and the objective is to optimize this response (Box and Draper 1987). RSM investigates the interaction between several input variables and one or more response variables. The first step of RSM is to define the limits of the experimental domain to be explored. The next step is the planning of accomplishing the experiments by means of RSM using either central composite design (CCD) or Box–Behnken design (BBD). RSM approach is the procedure for determining the relationship between various process parameters with the various machining criteria and for exploring the effect of these process parameters on the coupled responses which are in the form of a mathematical equation. Gunaraj and Murugan (1999) highlighted the use of RSM to develop mathematical models and plot contour graphs relating important input parameters to responses in submerged arc welding of pipes. The effect of the laser welding parameters on the bead geometry of AISI304 stainless steel has been investigated by Manonmani, Murugan, and Buvanasekaran (2007). In this study, the relationship between the process parameters (beam power, welding speed, and beam incidence angle) and the weld-bead parameters has been developed using RSM. Rajakumar, Muralidharan, and Balasubramanian (2011) optimized the process parameters of the friction stir welding process using RSM. Vasantharaja and Vasudevan (2018) carried out the optimization of the A-TIG welding process parameters to achieve the desired weld-bead geometrical parameters such as depth of penetration, bead width, and HAZ width by using the CCD technique of RSM. A second-order response surface model has been developed for correlating the process parameters with the weld-bead shape parameters. The authors also used the desirability approach in order to perform multi-objective optimization. The RSM technique has also been successfully used in combination of the Taguchi method to optimize the process parameters and develop input–output empirical relationships (Boulahem, Salem, and Bessrour 2015; Muhammad et al. 2013). The most important outcome of the RSM technique is also that it gives a mathematical relationship between the input and output parameters which can be fed to advanced meta-heuristics optimization techniques for a single- as well as multi-objective optimization.

### 4.2.5 Meta-Heuristics Methods

However, in the real manufacturing process, several conflicting responses (i.e., output responses) may affect the optimal solution because of the nonlinear characteristics of inputs with respect to the output responses. The objective function may be multimodal (i.e., more than one local minimum or maximum), while the main aim is to evaluate the global optimal values within the given search space/domain. Classical methods (viz. DOE, linear programming, dynamic programming, etc.) are found inefficient to handle these types of problems; therefore, advanced optimization algorithms (*heuristics and meta-heuristics methods*) are developed to seek for the feasible solution as they intend to find a solution near to global optimum in lesser time and with lesser computational effort. The use of meta-heuristics approach for the optimization of the welding process is gaining popularity. Welding process itself is a very complex phenomenon depending on a number of parameters. Also, the difficulty is even more as there are a number of conflicting parameters. For example, welding current is directly affecting the weld penetration as well as heat input. If we increase the welding current, the weld penetration will increase which is desirable. However, this will in turn increase in the heat input ultimately decreasing the mechanical properties that are undesirable. The use of this meta-heuristics optimization method solves this problem to some extent. The prerequisite for the optimization by this technique is a unique mathematical equation corresponding to each output response. Thus, several research studies have surfaced on optimization by using a combination of the Taguchi method, RSM, ANN, etc. as shown in Table 4.1. There are two widely used methods of optimization by using this optimization technique, i.e., weighted additive utility function (WAUF) technique and Pareto analysis. During the past two decades, researchers had developed some good-quality advanced optimization techniques such as GA, simulated annealing (SA), artificial bee colony (ABC), ant colony optimization (ACO), particle swarm optimization (PSO), teaching-learning-based optimization (TLBO), etc.; however, information about their implementation in the optimization of welding process parameters is limited in open literature.

Out of the available literature, the maximum attempt has been carried out with GA (Dave et al. 2016). The GA is a powerful, general-purpose optimization algorithm which uses the philosophy of "survival of the fittest" theory. GA was developed on the basis of probability that the global optimum for any objective is searched in a parallel and random manner through cross-over, mutation, and reproduction operations. The GA technique is widely used for optimization problems as it has several advantages over conventional methods of optimization which tends to get trapped at a local optimum value in search space. In this technique, two or more objectives are combined in a single objective by the weighted sum. In line with this, several researchers have attempted multi-objective optimization of input welding parameters of different welding processes with different optimization algorithms. In one of their studies, Muhammad et al. (2013) attempted to optimize the quality features for the resistance spot welding process. A combination of Taguchi method and RSM was used. Weld nugget and HAZ were simultaneously selected as the quality features and a mathematical model was developed for predicting the weld zone development. The developed models were well fitted and optimum parameters such as welding current, weld time, and hold time were obtained. Confirmations tests were carried out which were quite close to predicted values and hence the model was validated. Additionally, the contribution of each input factor on the output feature was also reported. Another study on small-scale resistance spot welding was carried out by Zhao et al. (2014). In this study, principle component analysis was carried out in which all quality features were converted into an independent quality characteristic. Regression equations were developed using RSM and the approach was found effective and feasible. Katherasan et al. (2014) used the ANN for developing the relationship between input and output variables for flux-cored arc welding process. Subsequently, the optimization was carried out by employing a PSO technique. The confirmation results indicated that ANN simulation using MATLAB® software was able to predict the bead profiles. The percentage error between predicted and actual output parameters was negligibly small. The same approach was selected by Ajith, Barik, et al. (2015) in their studies pertaining to multi-objective optimization of friction welding parameters for UNS S32205 steel. Four different welding parameters were selected for optimization and PSO technique was subsequently used. Confirmation tests were done subsequently using optimized parameters and the effects of input parameter on output variables were also assessed. Similarly, Sathiya et al. (2009) investigated the use of three different evolutionary optimization techniques such as PSO, GA, and SA for friction welding of stainless steels. It was reported that GA outperformed in predicting and optimizing the input parameters for friction welding. Confirmation tests were carried out for the optimized parameters and the model was validated. Optimization of A-TIG welding parameters was carried out for stainless steel type 316LN and 304LN by Vasudevan et al. (2007) as well as Chandrasekhar and Vasudevan (2010). In both studies, GA was used as the optimization method with use of regression models in former and use of ANN in later. WAUF was used to combine all responses as a single-objective function. In both studies, input parameters like welding current, voltage, arc gap, and torch speed were selected. The former study gave the regression equations to optimize and predict for achieving desired BW and DOP and reinforcement height (RH), whereas the later studies targeted only DOP and BW. It was reported from these studies that the GA technique suited fairly well for predicting and optimizing the A-TIG parameters. It was also reported that GA gave alternate parameters for achieving desired output parameters. Similar optimization and welding procedures were carried out by Subashini, Madhumitha, and Vasudevan (2012) as well as Vasudevan et al. (2010) but on Modified 9Cr-1Mo steel. However, an additional output parameter of HAZ width was also predicted and optimized. The GA code was found to be fairly suited for prediction and optimization of weld morphological dimensions. From the above literature review, it can be summarized that weld-bead shape parameters such as DOP, BW, and HAZ width are the most important parameters for getting the desired weld quality encompassing WAUF. In spite of the encouraging results, few concerns in using WAUF approach are the fact that proper selection of weights to different input parameters is necessary which is at the discretion of the designer.

Thus, the approach of generating Pareto optimal solutions for a given problem is gaining popularity in recent times. A Pareto optimal set is a set of solutions that are non-inferior with respect to each other (Deb 2001; Udayakumar et al. 2014). In other words, during simultaneous optimization of the variables, moving from one point on Pareto optimal set toward another, a certain amount of sacrifice is made in one objective to achieve some amount of increase in the other. This approach was used by Udayakumar et al. (2014) in their study related to multi-objective optimization of friction welding parameters for UNS 32760 steel joints. A Pareto optimal set of solutions was obtained for simultaneous optimization of two output parameters (corrosion resistance property and impact strength). The RSM technique was used to develop the mathematical models and further optimization was carried out by GA. Similarly, in the research by Ajith, Husain, et al. (2015) friction welding parameters were optimized by forming a Pareto set of solutions for weld hardness as well as the tensile strength of UNS 32760 steel joints. However, it is a well-known fact that apart from these, there can be other quality parameters which can be of interest and which dictates the weld quality. A Pareto optimal set of solutions covering all the major quality parameters shall serve as a guide to design and manufacturing engineers for deciding the input parameters during actual production. A typical Pareto front is as shown in Figure 4.2. The Pareto fronts are a trade-off between the conflicting parameters.

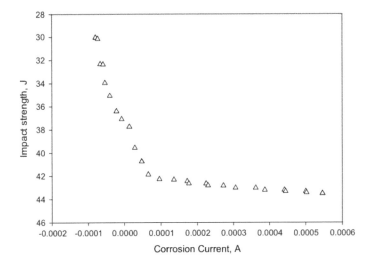

**FIGURE 4.2**
Pareto optimal front (Udayakumar et al. 2014). (With kind permission from Elsevier.)

## 4.3 Summary

This chapter provides the benefits of employing optimization strategies for the welding process. Optimization strategies are mainly divided into three broad categories, namely, Taguchi technique, RSM technique, and meta-heuristic optimization techniques.

In the different welding processes, several input process parameters have been selected/experimented by researchers to optimize different output response characteristics. The Taguchi method has several benefits which make the method favored by the experimental scientists particularly. The immediate benefit is the reduction in the number of experimental trials needed to analyze the effect of selected input parameters on output characteristics which involves unnecessary cost, time, and wastage of resources. The Taguchi method enables the researchers to contemplate a higher number of input parameters and analyze their effect on selected quality characteristics. This particular approach is certainly useful when a preliminary investigation is to be carried out on some completely new material where limited or no data are available regarding the response of this new material to a particular welding process. However, the major drawback of this method is the fact that Taguchi method in principle proposes an individual set of parameters for each output characteristics. In industrial application, there are several instances wherein, there is a need to optimize more than one output characteristic parameter. In such cases, the Taguchi method is certain times incapable of providing the solution. However, several approaches such as GRA can be used for the multi-objective optimization. In addition to this, the Taguchi method has proved effective in providing data for the development of mathematical models by ANN, SAA, and multi-objective optimization methods.

The RSM approach can be considered a step ahead of the Taguchi method as they provide a mathematical relationship between input and output parameters. RSM is a collection of mathematical and statistical techniques that are useful for the modeling and analysis of problems in which output or the response is influenced by several input variables and the objective is to optimize this response. RSM also reduces the number of experimental trials by using either CCD or BBD approach. The research survey showed that RSM determines the relationship between various process parameters with the various machining criteria. However, RSM technique gives the set of the optimum level of input parameters without considering the constraints. Hence, an optimum parameter setting with constraints is the requirement for the process which can satisfy all such conflicting objectives at the same time. Such situations can be tackled conveniently by making use of meta-heuristic optimization techniques for the parameters optimization. The relations between the input and output of the RSM method are used as an input data to be fed to meta-heuristics methods to simultaneously optimize more than one output parameters.

The meta-heuristics methods are generally complete packages which can effectively be used to predict and optimize single- as well as multi-objective optimization problems. The WAUF method has been widely used for the optimization by forming an objective function giving weights to different input responses of the welding processes. However, a better approach was developed by the researchers termed as Pareto fronts by modifying the existing techniques to suit for multi-objective optimization of process parameters, particularly for conflicting parameters. These Pareto fronts basically present a trade-off between two conflicting parameters and the manufacturers can select any of these points on the fronts. A targeted output response can be achieved by this methodology.

## 4.4 Future Scope

Welding will be one of the most important and influential industrial activities. However, in the current scenario, every manufacturing operation including welding is shifting toward sustainability, wherein apart from the required output several aspects the environmental impact shall also be reduced. The optimization of welding process parameters will hence gain popularity as it optimizes the process parameters and fine-tunes it to get maximum output with minimum input. Several materials vastly used in important applications are yet to be optimized and this can be an upcoming activity. Several novel meta-heuristics such as TLBO, heat transfer search (HTS), and passive vehicle search (PVS) can also be used for optimization of welding processes.

## References

Abhulimen, IU, and JI Achebo. 2014. "Application of artificial neural network in predicting the weld quality of a tungsten inert gas welded mild steel pipe joint." *International Journal of Scientific & Technology Research* 3 (1):277–285.

Ajith, PM, BK Barik, P Sathiya, and S Aravindan. 2015. "Multiobjective optimization of friction welding of UNS S32205 duplex stainless steel." *Defence Technology* 11 (2):157–165.

Ajith, PM, TMA Husain, P Sathiya, and S Aravindan. 2015. "Multi-objective optimization of continuous drive friction welding process parameters using response surface methodology with intelligent optimization algorithm." *Journal of Iron and Steel Research, International* 22 (10):954–960.

Anand, K, BK Barik, K Tamilmannan, and P Sathiya. 2015. "Artificial neural network modeling studies to predict the friction welding process parameters of Incoloy 800H joints." *Engineering Science and Technology, an International Journal* 18 (3):394–407.

Asif, MM, KA Shrikrishna, and P Sathiya. 2016. "Optimization of process parameters of friction welding of UNS S31803 duplex stainless steels joints." *Advances in Manufacturing* 4 (1):55–65.

Bhattacharya, A, and S Singla. 2017. "Mechanical properties and metallurgical characterization of dissimilar welded joints between AISI 316 and AISI 4340." *Transactions of the Indian Institute of Metals* 70 (4):893–901.

Boulahem, K, SB Salem, and J Bessrour. 2015. "Surface roughness model and parametric welding optimization in friction stir welded AA2017 using Taguchi method and response surface methodology." In M. Chouchane, T. Fakhfakh, H. B. Daly, N. Aifaoui, F. Chaari (eds.). *Design and Modeling of Mechanical Systems-II*, 83–93. Springer, Berlin.

Box, GEP, and NR Draper. 1987. *Empirical Model-Building and Response Surfaces.* John Wiley & Sons, New York.

Chandrasekhar, N, and M Vasudevan. 2010. "Intelligent modeling for optimization of A-TIG welding process." *Materials and Manufacturing Processes* 25 (11):1341–1350.

Correia, DS, CV Gonçalves, SS da Cunha Jr, and VA Ferraresi. 2005. "Comparison between genetic algorithms and response surface methodology in GMAW welding optimization." *Journal of Materials Processing Technology* 160 (1):70–76.

Dave, S, J Vora, N Thakkar, A Singh, S Srivastava, B Gadhvi, V Patel, and A Kumar. 2016. "Optimization of EDM drilling parameters for Aluminum 2024 alloy using Response Surface Methodology and Genetic Algorithm." *Key Engineering Materials.* doi:10.4028/www.scientific.net/KEM.706.

Deb, K. 2001. *Multi-Objective Optimization Using Evolutionary Algorithms.* John Wiley & Sons Inc., New York.

Deepandurai, K, and R Parameshwaran. 2016. "Multiresponse optimization of FSW parameters for cast AA7075/SiCp composite." *Materials and Manufacturing Processes* 31 (10):1333–1341.

Dhas, JER, and SJH Dhas. 2012. "A review on optimization of welding process." *Procedia Engineering* 38:544–554.

Dhas, JER, and S Kumanan. 2011. "Optimization of parameters of submerged arc weld using non conventional techniques." *Applied Soft Computing* 11 (8):5198–5204.

Fuzeau, J, M Vasudevan, and V Maduraimuthu. 2016. "Optimization of welding process parameters for reduced activation ferritic-martensitic (RAFM) steel." *Transactions of the Indian Institute of Metals* 69 (8):1493–1499.

Gunaraj, V, and N Murugan. 1999. "Application of response surface methodology for predicting weld bead quality in submerged arc welding of pipes." *Journal of Materials Processing Technology* 88 (1–3):266–275.

Hakan, A, A Bayram, U Esme, Y Kazancoglu, and O Guven. 2010. "Application of grey relation analysis (GRA) and Taguchi method for the parametric optimization of friction stir welding (FSW) process." *Materials and Technology* 44:205.

Hsiao, YF, YS Tarng, and WJ Huang. 2007. "Optimization of plasma arc welding parameters by using the Taguchi method with the grey relational analysis." *Materials and Manufacturing Processes* 23 (1):51–58.

Huang, H-Y. 2010. "Effects of activating flux on the welded joint characteristics in gas metal arc welding." *Materials & Design (1980–2015)* 31 (5):2488–2495.

Juang, SC, and YS Tarng. 2002. "Process parameter selection for optimizing the weld pool geometry in the tungsten inert gas welding of stainless steel." *Journal of Materials Processing Technology* 122 (1):33–37.

Kangazian, J, and M Shamanian. 2016. "Multiresponse optimization of pulsed-current gas tungsten arc welding (PCGTAW) for AISI 304 stainless steel to St 52 steel dissimilar welds." *Metallography, Microstructure, and Analysis* 5 (3):241–250.

Kasman, Ş. 2013. "Multi-response optimization using the Taguchi-based grey relational analysis: a case study for dissimilar friction stir butt welding of AA6082-T6/AA5754-H111." *The International Journal of Advanced Manufacturing Technology* 68 (1–4):795–804.

Katherasan, D, JV Elias, P Sathiya, and AN Haq. 2013. "Modeling and optimization of flux cored arc welding by genetic algorithm and simulated annealing algorithm." *Multidiscipline Modeling in Materials and Structures* 9 (3):307–326. doi: 10.1108/MMMS-03-2012-0008.

Katherasan, D, JV Elias, P Sathiya, and AN Haq. 2014. "Simulation and parameter optimization of flux cored arc welding using artificial neural network and particle swarm optimization algorithm." *Journal of Intelligent Manufacturing* 25 (1):67–76.

Kiaee, N, and M Aghaie-Khafri. 2014. "Optimization of gas tungsten arc welding process by response surface methodology." *Materials & Design (1980–2015)* 54:25–31.

Kumar, A, and S Sundarrajan. 2009. "Optimization of pulsed TIG welding process parameters on mechanical properties of AA 5456 aluminum alloy weldments." *Materials & Design* 30 (4):1288–1297.

Lin, H-L. 2013. "Optimization of Inconel 718 alloy welds in an activated GTA welding via Taguchi method, gray relational analysis, and a neural network." *The International Journal of Advanced Manufacturing Technology* 67 (1–4):939–950.

Lin, H-L, and C-P Chou. 2010. "Optimization of the GTA welding process using a combination of the Taguchi method and a neural-genetic approach." *Materials and Manufacturing Processes* 25 (7):631–636.

Lin, H-L, and J-C Yan. 2014. "Optimization of weld bead geometry in the activated GMA welding process via a grey-based Taguchi method." *Journal of Mechanical Science and Technology* 28 (8):3249–3254.

Lockhart, SD, and CM Johnson. 2000. *Engineering Design Communication: Conveying Design through Graphics.* Prentice Hall, Upper Saddle River, NJ.

Magudeeswaran, G, SR Nair, L Sundar, and N Harikannan. 2014. "Optimization of process parameters of the activated tungsten inert gas welding for aspect ratio of UNS S32205 duplex stainless steel welds." *Defence Technology* 10 (3):251–260.

Manonmani, K, N Murugan, and G Buvanasekaran. 2007. "Effects of process parameters on the bead geometry of laser beam butt welded stainless steel sheets." *The International Journal of Advanced Manufacturing Technology* 32 (11–12):1125–1133.

Muhammad, N, YHP Manurung, R Jaafar, SK Abas, G Tham, and E Haruman. 2013. "Model development for quality features of resistance spot welding using multi-objective Taguchi method and response surface methodology." *Journal of Intelligent Manufacturing* 24 (6):1175–1183.

Nagaraju, S, P Vasantharaja, N Chandrasekhar, M Vasudevan, and T Jayakumar. 2016. "Optimization of welding process parameters for 9Cr-1Mo steel using RSM and GA." *Materials and Manufacturing Processes* 31 (3):319–327.

Narwadkar, A, and S Bhosle. 2016. "Optimization of MIG welding parameters to control the angular distortion in Fe410WA steel." *Materials and Manufacturing Processes* 31 (16):2158–2164.

Palani, PK, and N Murugan. 2007. "Optimization of weld bead geometry for stainless steel claddings deposited by FCAW." *Journal of Materials Processing Technology* 190 (1–3):291–299.

Pujari Srinivasa Rao, PR, CKR Rao, and SP Priyangeli. 2016. "Effect of halide fluxes and welding parameters on penetration and haz of titanium alloy." *Journal of Production Engineering* 19 (1):7.

Rajakumar, S, C Muralidharan, and V Balasubramanian. 2011. "Response surfaces and sensitivity analysis for friction stir welded AA6061-T6 aluminium alloy joints." *International Journal of Manufacturing Research* 6 (3):215–235.

Rao, RV, and VD Kalyankar. 2013. "Experimental investigation on submerged arc welding of Cr–Mo–V steel." *The International Journal of Advanced Manufacturing Technology* 69 (1–4):93–106.

Sahu, PK, and S Pal. 2015. "Multi-response optimization of process parameters in friction stir welded AM20 magnesium alloy by Taguchi grey relational analysis." *Journal of Magnesium and Alloys* 3 (1):36–46.

Santana, LM, UFH Suhuddin, MH Ölscher, TR Strohaecker, and JF dos Santos. 2017. "Process optimization and microstructure analysis in refill friction stir spot welding of 3-mm-thick Al-Mg-Si aluminum alloy." *The International Journal of Advanced Manufacturing Technology* 92 (9–12):4213–4220.

Sathiya, P, PM Ajith, and R Soundararajan. 2013. "Genetic algorithm based optimization of the process parameters for gas metal arc welding of AISI 904 L stainless steel." *Journal of Mechanical Science and Technology* 27 (8):2457–2465.

Sathiya, P, S Aravindan, AN Haq, and K Paneerselvam. 2009. "Optimization of friction welding parameters using evolutionary computational techniques." *Journal of Materials Processing Technology* 209 (5):2576–2584.

Sathiya, P, MYA Jaleel, and D Katherasan. 2010. "Optimization of welding parameters for laser bead-on-plate welding using Taguchi method." *Production Engineering* 4 (5):465–476.

Sathiya, P, MYA Jaleel, D Katherasan, and B Shanmugarajan. 2011. "Optimization of laser butt welding parameters with multiple performance characteristics." *Optics & Laser Technology* 43 (3):660–673.

Subashini, L, P Madhumitha, and M Vasudevan. 2012. "Optimisation of welding process for modified 9Cr-1Mo steel using genetic algorithm." *International Journal of Computational Materials Science and Surface Engineering* 5 (1):1–15.

Udayakumar, T, K Raja, TM Afsal Husain, and P Sathiya. 2014. "Prediction and optimization of friction welding parameters for super duplex stainless steel (UNS S32760) joints." *Materials & Design* 53:226–235.

Vasantharaja, P, and M Vasudevan. 2018. "Optimization of A-TIG welding process parameters for RAFM steel using response surface methodology." *Proceedings of the Institution of Mechanical Engineers, Part L: Journal of Materials: Design and Applications* 232 (2):121–136.

Vasudevan, M, V Arunkumar, N Chandrasekhar, and V Maduraimuthu. 2010. "Genetic algorithm for optimisation of A-TIG welding process for modified 9Cr–1Mo steel." *Science and Technology of Welding and Joining* 15 (2):117–123.

Vasudevan, M, AK Bhaduri, B Raj, and KP Rao. 2007. "Genetic-algorithm-based computational models for optimizing the process parameters of A-TIG welding to achieve target bead geometry in type 304 L (N) and 316 L (N) stainless steels." *Materials and Manufacturing Processes* 22 (5):641–649.

Vijayan, S, Raju R, Rao SRK, and JJ Vora. 2010. "Multiobjective optimization of friction stir welding process parameters on aluminum alloy AA 5083 using Taguchi-based grey relation analysis." *Materials and Manufacturing Processes* 25 (11):1206–1212.

Wang, Y, and DO Northwood. 2008. "Optimization of the polypyrrole-coating parameters for proton exchange membrane fuel cell bipolar plates using the Taguchi method." *Journal of Power Sources* 185 (1):226–232.

Xu, WH, SB Lin, CL Fan, and CL Yang. 2015. "Prediction and optimization of weld bead geometry in oscillating arc narrow gap all-position GMA welding." *The International Journal of Advanced Manufacturing Technology* 79 (1–4):183–196.

Yang, LJ, MJ Bibby, and RS Chandel. 1993. "Linear regression equations for modeling the submerged-arc welding process." *Journal of Materials Processing Technology* 39 (1–2):33–42.

Yang, LJ, RS Chandel, and MJ Bibby. 1993. "An analysis of curvilinear regression equations for modeling the submerged-arc welding process." *Journal of Materials Processing Technology* 37 (1–4):601–611.

Zhao, D, Y Wang, S Sheng, and Z Lin. 2014. "Multi-objective optimal design of small scale resistance spot welding process with principal component analysis and response surface methodology." *Journal of Intelligent Manufacturing* 25 (6):1335–1348.

# 5

## Additive Manufacturing with Welding

**Manish Kumar, Abhay Sharma,**
**Uttam Kumar Mohanty, and S. Surya Kumar**
*Indian Institute of Technology Hyderabad*

### CONTENTS

5.1   Introduction ................................................................................. 77
5.2   Welding-Based Additive Manufacturing Techniques ........................... 79
      5.2.1   Powder-Based AM Technology ............................................. 79
      5.2.2   Wire-Feed-Based WAM .................................................... 83
5.3   Practical Considerations in WAM .................................................. 84
      5.3.1   Path Planning ................................................................ 84
      5.3.2   Overhanging Feature in WAM ............................................ 86
5.4   Recent Developments in the Fabrication of Parts by WAM Process .... 92
      5.4.1   Advance Power Sources .................................................... 92
      5.4.2   In-Situ Sensing and Adaptive Process Control ..................... 93
      5.4.3   Hybrid WAM Processes .................................................... 94
      5.4.4   Rolling-Assisted WAM ..................................................... 94
5.5   Summary and Conclusion .............................................................. 97
References ........................................................................................... 98

## 5.1 Introduction

The additive manufacturing (AM) is one of the latest additions to manufacturing processes wherein parts are created by digitally controlled layer-by-layer deposition and bonding. It is a near-net shape (NNS) process that aims to create semifinished products as close as possible to the actual geometry with the objective to maximize raw material utilization and a minimization of number of finishing steps (e.g., machining operations). The AM is a promising technology that can fabricate complex geometry components with less human intervention and high material efficiency. According to American Society for Testing and Materials, AM is "a process of joining materials to make objects from 3D model data, usually layer upon layer, as opposed to subtractive manufacturing methodologies" (A. S. T. M. Standard F2792 2012). The AM has the ability to fabricate a single-component structure with complex shapes, which is very challenging for other traditional manufacturing processes. Intricate and complex geometries can be manufactured using AM with less post-processing. The AM offers increased design freedom, which enables the engineers and designers to fabricate unique components in low volumes in an economical way. It is also a step ahead from rapid prototyping as its primary goal is to manufacture directly fully functional parts from digital models using actual materials, opposite to rapid prototyping that fabricates non-functional parts (Hascoet et al. 2014). The AM technologies and methods grabbed the attention of various sectors such as automotive, medical, and aerospace. Figure 5.1 shows the percentage implementation of AM in various sectors.

The AM involves the creation of a 3D model, slicing the model into cross-sectional areas, and planning of path for filling the area. Being an automated process, the AM has the following advantages over conventional manufacturing technologies through which the importance of AM can be realized:

  i. *Material efficiency.* AM processes have high material efficiency due to its ability to fabricate NNS of the component in a layer-by-layer manner. Unlike conventional machining or material removal processes where a large amount of material needs to be removed, the AM mostly requires finishing operations.

 ii. *Resource efficiency.* AM technology needs limited additional resources such as jigs, fixtures, cutting tools, and coolants that are essential for most conventional subtractive manufacturing. Therefore, parts can be made by small manufacturers that are close to customers, which results in improved supply chain dynamics.

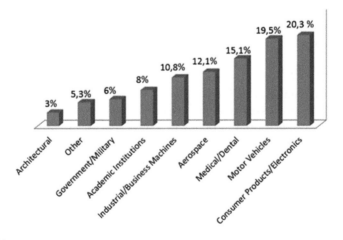

**FIGURE 5.1**
Sector-wise implementation of AM process (Bikas et al. 2016). [With the kind permission of Springer, https://link.springer.com/article/10.1007/S00170-015-7576-2 (Open access).]

iii. *Part flexibility.* There is no need to sacrifice part functionality for the ease of manufacturing as the parts with complex geometry are possible in a single piece.

iv. *Production flexibility.* The additional resources like tool and component support systems in the AM are very limited that results in economic small batch production. Problems of line balancing and production bottlenecks are eliminated because complex parts are produced in a single piece. The quality of the parts depends on the process instead of operator skills due to which products can be modified easily according to customer demand.

v. *Environmental benefit.* The AM is environment-friendly due to no material wastage and no tooling, making it a promising technology in all aspects. The subtractive process dominated until the late 1990s wherein buy-to-fly (BTF) ratio (mass ratio of material used and the final product) is 5:1, but in some cases, it goes up to 20:1 (Yilmaz and Ugla 2016).

The AM with polymers is currently mature enough for the production of small and large parts (with the longest axis length at minimum 1–2 m). The metal AM is an emerging technology with which complex components can be manufactured. This chapter focuses on welding-based additive manufacturing (WAM) that is the core of metal additive manufacturing (MAM).

## 5.2 Welding-Based Additive Manufacturing Techniques

As the fabrication of the metallic AM components necessities localized application of a heat source, welding becomes an ideal tool for AM. The raw material in the form of metal powder or wire is melted or sintered with high-intensity heat sources, such as laser, electron beam, and welding arc. Some of the complex metallic components produced by WAM processes like wire and arc additive manufacturing (WAAM), laser additive manufacturing (LAM), and electron beam additive manufacturing (EBAM) are shown in Figure 5.2.

The WAM can be classified based on the motion controller, raw material, and heat source as shown in Figure 5.3. The way the raw material is used (i.e., in powder or wire form) is the primary attribute of the AM technology that is supported by the heat source and the motion control.

### 5.2.1 Powder-Based AM Technology

The powder-based AM is of two types—a powder bed and a powder feed. In the former, the powder is spread in the powder bed system and then

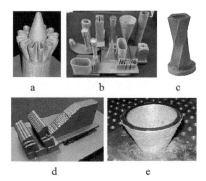

**FIGURE 5.2**
AM parts fabricated using laser-based powder bed/feed (a and b), electron beam based (c), laser-based wire feed (d), and arc based (e) (Karunakaran et al. 2010). (With the kind permission of Elsevier.)

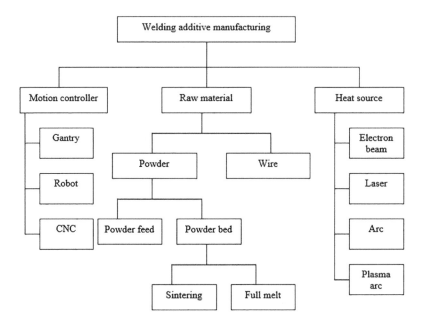

**FIGURE 5.3**
Classification of welding-based additive manufacturing.

selectively melted, while in the latter, the powder is fed in the presence of a heat source that melts the metal and deposits. The powder bed AM, also known as powder bed fusion (PBF), consists of a roller or a blade to spread powder to form a layer of thickness up to 20–100 µm depending upon the process as well as the particle size of the powder (Figure 5.4) (Dongdong 2015). A focused energy source (i.e., laser beam, electron beam, or plasma arc) is used to selectively fuse the grains of the powder. The energy source moves

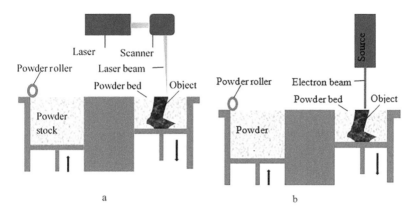

**FIGURE 5.4**
Powder bed additive manufacturing: (a) laser based and (b) electron beam based.

as per the path planned for the deposition of the layer. The fused powder forms the component, and the unfused powder supports the component. The unfused powder is cleaned away and reused after the completion of the fabrication process. The PBF technique is suitable for manufacturing highly complex parts containing undercuts because the powder acts as a self-supporting structure. Furthermore, the powder bed system is classified based on the ways of powder melting (i.e., fully melting and partially melting).

The commonly used powder bed systems are SLS (selective laser sintering), SLM (selective laser melting), and EBM (electron beam melting). Figure 5.4a and b shows laser powder bed and electron powder bed systems, respectively. These systems are used to fabricate complex functional parts such as turbine blades, fuel injectors, heat exchangers, dental bridges, medical implant, micro-turbines, wheel suspensions, brake disks in the automotive industry, and jewelry. The SLS is one kind of PBF process in which laser is used to sinter the powder bed particle. The temperature of the sintered material is a few degrees below the melting point temperature of the powder particle. The working principle of SLM is similar to SLS, but the complete melting of powder particle takes place. The SLM process uses an inert gas, and a higher power laser beam and often faster than SLS. The EBM-based AM is a rapidly growing and comparatively new process, which is similar to SLM, but it can fabricate only metallic objects. A high-power electron beam melts the powder layer in a vacuum chamber to fabricate the object. The non-liquefied powder is cleared away after curing of the hot surface. The EBM process offers higher-throughput and more uniform temperature distribution during fusion, which provides excellent strength in comparison to the SLM process. The process is used in the manufacturing of high standard parts of airplanes and medical applications as it allows high quality and finishes (Loughborough University, AMRG Group). The EBM is successfully used to fabricate component of metals that have a high affinity for oxygen.

Among the three versions, the SLS is relatively slow and has an issue of non-uniform temperature distribution that may lead to thermal distortion and cracks on the component. Despite that, SLS has a high degree of accuracy and surface quality. Hence, it is widely used metal AM processes (Dongdong 2015) to fabricate porous material. One of the noticeable examples of powder-based AM—first wall (FW) panel part of the international thermonuclear experimental reactor (ITER) that is a very complex design—is fabricated using SLM- and EB-based AM processes (Zhong et al. 2017).

The powder feed AM includes several variants, such as laser engineering net shaping (LENS), laser powder deposition (LPD), direct metal deposition (DMD), and selective laser cladding (SLC). A typical LENS system is shown in Figure 5.5, wherein the powder nozzle and the head of the laser travel as an integral entity. The powder is injected/blown by gravity or a pressurized gas into a laser-focused beam to create a molten pool. The component is fabricated by depositing the material layer by layer. It is also used to repair parts but still needs post-production treatments. The LENS can fabricate substantially complex and larger parts, even up to several meters long in comparison to powder bed, but the surface quality and the accuracy are lower in this case (Gibson et al. 2010). The powder and air stream are introduced directly into the focus point of the laser beam on the substrate in LPD to form the layer of parts. In the DMD process, the powder is injected by nozzles (typically three in number) instead of one to focus directly on the processing head.

The low yield and deposition rate of powder-based AM limits it to manufacture small components. The process is expensive and time-consuming for the manufacturing of large component due to the problems associated with powder recycling, contamination, and storage. The fabricated components

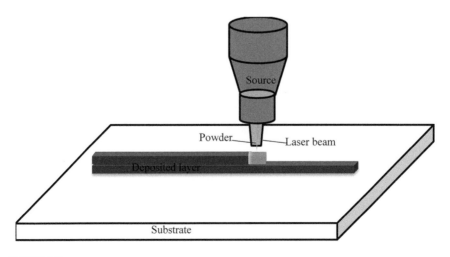

**FIGURE 5.5**
Powder feed system.

may also suffer from high surface roughness and residual gas porosity. The majority of these concerns, especially component size can be obviated using metal wire as an alternative feedstock material.

### 5.2.2 Wire-Feed-Based WAM

Wire-feed WAM comprises a wire feeder, a heat source, and a motion control unit. In this process, the wire melts and creates a weld pool by a heat source (e.g., laser, electron beam, plasma, and arc) to manufacture the components. The material usage efficiency of wire-feed based WAM is very high and even goes up to 100% without exposing operators to the hazardous powder environment that makes it environment-friendly (Ding et al. 2015a). The wire-feed-based WAM processes are designated based on the heat sources [e.g., EBF3 (electron beam freeform fabrication), WLAM (wire and laser additive manufacturing), WAAM (wire arc additive manufacturing)]. Laser, electron beam, and plasma are the energy sources primarily used to fabricate small component due to very low deposition rate and energy efficiency, whereas arc welding [Gas tungsten arc welding (GTAW) or gas metal arc welding (GMAW)] is used for producing comparatively large components with moderate complexity. A schematic diagram of the WLAM, EBF3, and WAAM process is shown in Figure 5.6.

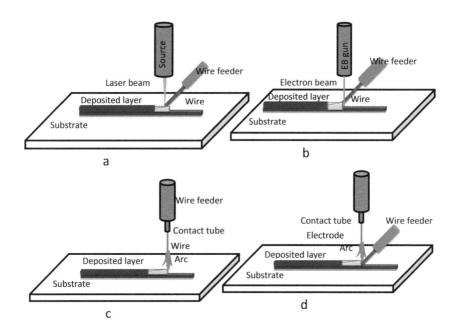

**FIGURE 5.6**
Wire-feed additive manufacturing: (a) laser based, (b) electron beam based, (c) arc based (GMAW), and (d) arc based (GTAW).

Usually, the deposition rate for laser or electron beam is in the range of 2–10 g/min, whereas the same is 50–130 g/min for the arc process. The equipment cost of arc welding is also low as compared to the electron beam and laser. Wire-feed WLAM has a very good "precision" but has a very low energy efficiency (i.e., 2%–5%), whereas EBF3 has slightly high (i.e., 15%–20%) but requires a high vacuum. EBF3 is suitable to fabricate component of materials having a high affinity for oxygen (e.g., Titanium). The energy efficiency of GMAW or GTAW even goes up to 90% in some circumstances (Ding et al. 2015b). The GMAW is a well-established process in joining of two components and now becoming popular in the domain of WAM fabrication because of its promising energy and material efficiency. In GMAW process, the arc is created between a consumable wire electrode and the metal workpiece to melt the wire and to create a weld pool for forming a bead on cooling. Cold metal transfer (CMT) is an advanced GMAW variant in which back and forth motion of the wire makes controlled short-circuit transfer without spattering and less heat input (Ding et al. 2015a). Selection of AM processes is a trade-off between resolution and deposition rate for a specific part. The wire-feed AM is more efficient due to high material deposition, but less accurate as compared to powder-based AM. The fabrication of the component by the WAM may consist of a single-pass multilayer or multi-pass multilayer. The main concern in single-pass multilayer parts is geometric accuracy, whereas in the multi-pass multilayer parts, thermal aspect is also critical along with the tool-path planning.

## 5.3 Practical Considerations in WAM

### 5.3.1 Path Planning

The path planning is primarily being able to produce sensible paths automatically for depositing various geometries with minimum residual stress and distortion (Ding et al. 2016a). The layers can be deposited through a number of tool paths that is a single start and stop for continuous material deposition. The geometrical feature, thermal effects, and process capabilities limitation are essential aspects need to be considered in the decision-making of path planning. Various tool-path planning approaches have been developed in WAM processes over the years to meet the requirements; some of them are shown in Figure 5.7.

The raster tool path (Figure 5.7a) fills the area line by line along one particular direction. It starts and frequently stops due to the presence of discontinuous paths, which is undesirable for WAAM because of instability during the start and end of the process. The raster produces a nonuniform deposition of material that has no resemblance with the ideal bead obtained in the

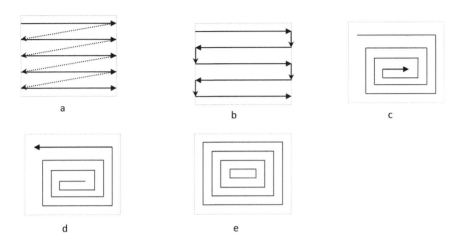

**FIGURE 5.7**
Typical path planning of additive manufacturing: (a) raster, (b) zigzag, (c) spiral in, (d) spiral out, and (e) contour.

steady-state condition. Zigzag tool-path generation is applied to overcome this problem by combining the separate parallel lines into a single continuous pass (Figure 5.7b). However, the discretization errors in both raster and zigzag may lead to accuracy reduction on the boundary of the component, if the tool motion direction is not parallel to edge. The accumulation of heat also occurs in certain regions due to the involvement of highly convoluted paths in zigzag. The spiral tool-path (Figure 5.7c,d) is another technique used to resolve the zigzag path complications in the WAM process, but it is appropriate for only particular geometrical models. The contour tool path (Figure 5.7e) offsets parallel to a boundary of the component recursively and, thus, can effectively address the issue of a geometrical feature in raster and zigzag. It reduces the distortion as well as anisotropy by altering the path direction regularly along the boundary (Jin et al. 2013). It is often preferred in wire-feed AM deposition over raster and zigzag in constant wall thickness.

The limitations of the aforementioned tool paths for complex geometries are overcome by algorithmic path plans such as the medial axis transformation (MAT) and adaptive path planning.

The medial axis is the locus of the centers of circles (spheres) of the different radius that touch or are close to at least one point on the surface of the object. The contour and MAT paths have constant step-over distances that may lead to an unfilled or overfilled area as shown in Figure 5.8. The contour path plan (Figure 5.8a) generates several closed paths by offsetting the boundary curves toward its interior. In the case of the WAAM, where step-over distance is large, the contour path may leave a gap as shown in Figure 5.8b. The MAT path (Figure 5.8c) generates void-free deposition as shown in Figure 5.8d, but it deposits extra material at the boundary. Post-processing like machining

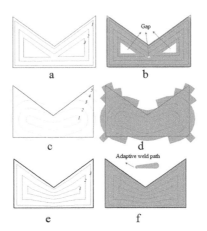

**FIGURE 5.8**
Effect of contour (a and b), MAT (c and d), and adaptive (e and f) path planning (Ding et al. 2016b). (With the kind permission of Elsevier.)

removes these extra materials for enhancing the accuracy at the cost of material and energy wastage. The limitations of contour path and MAT path are overcome by adaptive path planning wherein the voids and excess material is avoided by continuously varying step-over distances (Figure 5.8e) through of adjustment of the process parameters (Ding et al. 2016b), as shown in Figure 5.8f. The WAAM is specially equipped for adaptive path planning as different width within a layer can be achieved by changing welding speed and the wire feed.

## 5.3.2 Overhanging Feature in WAM

The overhanging features are indispensable in AM. Unlike the polymer 3D printing, the detachable support structures are not always possible in WAM. Some of the overhanging features in WAM parts are shown in Figure 5.9. Without any support from the bottom side during fabrication, the molten metal may cross flow and deform. In order to prevent deformation, the support structures are fixed to the fabricating platform through which excess

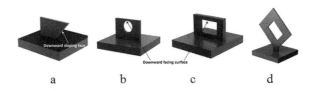

**FIGURE 5.9**
Overhanging structures: (a) downward sloping face, (b and c) downward-facing surfaces, and (d) downward sloping faces obtained by orientation in the building platform (Calignano 2014). (With the kind permission of Elsevier.)

heat conducted away from the component. The presence of the support structures increases the difficulty of post-processing operations. Therefore, the geometrical design and optimization of the support structures are essential to improve the sustainability and efficiency of a metallic object fabricated by the AM (Calignano 2014).

The overhanging features cannot always be avoided but can be minimized only by selecting an appropriate orientation. The feature of the part either is an overhanging or not, is determined by the orientation of the component during manufacturing. The surface area of the support structure depends on the part's built orientation. The best orientation in the manufacturing of a component with minimal support is selected from a list of the candidate orientations. In case, the support structures require for two orientations are equal, the orientation having a lower center of mass is selected.

In powder-bed-based AM process, the melt pool is created by the heat source that is supported by a bed of powder particles. The heat conducted by two different zones (i.e., solid supported and powder supported) wherein one has higher heat conduction rate while other much lower, respectively. This makes melt pool when supported by powder, large and deep, which sink into the powder due to gravitational and capillary forces. The melting penetration of melt into the powder bed is caused due to the Rayleigh–Taylor instability in the gravity (Chivel and Smurov 2011). The heat of the melt pool causes unintentional sintering of the supporting powder particles to form dross, which lowers the dimensional accuracy. The dross formation is unavoidable in the fabrication of the overhanging surface during PBF. The dross can be minimized by selecting proper process conditions through investigating the thermo-physical mechanism using numerical simulation— e.g., both high and low laser volume energy densities result in an inferior quality of downward-facing surface (Chen et al. 2017). The relatively smooth downward-facing surface is obtained at the optimal processing parameter due to the sound melt pool dimension and steady melt flow behavior.

Warping (Figure 5.10) is another effect that is caused due to the difference in thermal conductivity, as the thermal conductivity of a powder bed is approximately 100 times lower than a bulk solid (Rombouts et al. 2005). Therefore, support structures are often required to aid heat dissipation, especially at overhanging features. The finite-element method (FEM) is used to simulate the thermal induced deformation for different overhanging support patterns such as overhang length, the support column and the gap between solid pieces. The longer overhang length causes a more extensive deformation, whereas even a thin support column can prevent overhang curling (Cheng and Chou 2015).

The support cannot be avoided and its utilization becomes mandatory in some cases. The removal of support during the post-processing phase is a tedious task for delicate parts, may cause the breakdown of a small piece, and adversely affects the surface quality, which partially limits the freedom of design. Therefore, the focus in the powder bed AM is to reduce the

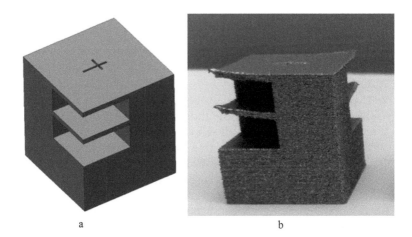

a b

**FIGURE 5.10**
Overhanging features: (a) CAD model and (b) warping of an overhanging feature in the fabrication of powder bed AM (Cheng and Chou 2015). (With the kind permission of Elsevier.)

number of supporting surfaces for improving the sustainability and the process efficiency without compromising the quality by the optimization and geometrical design of the support structures.

The conventional support has straight rectangular solid walls in the form of a grid of $x$ and $y$ lines, which are separated at a certain distance ($x$ hatching and $y$ hatching) or blocks that should be strong enough to bear both the vertical weight and other horizontal instabilities (Venuvinod and Ma 2013). The selection of hatching depends on the surface area. The problem associated with this kind of support is trapping of the loose powder inside the support structures during the fabrication that becomes unsuitable to recover; especially trapped raw powders washes away while removal by EDM-wire cutting (Hussein et al. 2013).

The support structures are not used in the manufacturing of the overhang features in wire-feed AM. Torch orientation and table orientation in following manners facilitate overhang features fabrication: (a) flat table and vertical torch position, (b) flat table and inclined torch position, (c) rotating table and vertical torch position, and (d) rotating table and inclined torch position, which are shown in Figure 5.11.

In the case of flat table and vertical torch position, the deposition of inclined parts is very difficult to control. The amount of inclination is decided by the horizontal step-over (i.e., offset) of the bead, which is affected by various welding process parameters. The angle of inclination decreases with increasing the step-over at constant welding speed, as shown in Figure 5.12a. The offset distance can be increased until the sum of the droplet impact, arc force, and the gravity is lesser than the surface tension. The angle of inclination to horizontal increases by increasing the wire-feed rate (Xiong et al. 2017), as shown in Figure 5.12b.

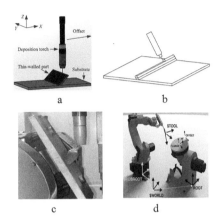

**FIGURE 5.11**
Method for fabricating overhanging features: (a) flat table and vertical torch (Xiong et al. 2017), (b) flat table and inclined torch, (c) rotational table and vertical torch (Panchagnula and Simhambhatla 2016), and (d) rotational table and inclined torch position (Ding et al. 2017). (With the kind permission of (a and d) Elsevier and (c) Springer.)

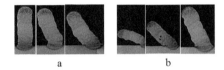

**FIGURE 5.12**
Effect of (a) offset distance 0, 0.5, and 1.0 mm and (b) wire feed and offset distance of 2.12 m/min and 1.32 mm, 2.92 m/min and 1.12 mm, and 4.67 m/min and 0.84 mm (Xiong et al. 2017). (With the kind permission of Elsevier.)

The last three approaches (Figure 5.11b–d) can accomplish a random angle inclination to fabricate components with a better quality, mainly via the rotating table and inclined torch position approach but require complicated mechanical system. The positioner or a turntable is required for tilting of the base, whereas a robot is required for the inclination of the torch position.

The inclined wall can also be built with a flat table and inclined torch at an angle to the substrate as shown in Figure 5.13a. The deposition of the first two or three beads is very critical in the generation of the inclined wall as hump forms in the region close to the substrate. To eliminate this effect, first, three layers are deposited by keeping torch to a vertical position with respect to the substrate, followed by two and one beads in the second and third layers, respectively, that generates a pyramidal structure with respect to the wall base. The acute angle of inclination in the fabrication of the inclined wall acts as a stress riser and get worse with smaller angles. Horizontal wall (i.e., full overhanging) can also be fabricated with the inclined torch method, as shown in Figure 5.13b (0°—collinear with the plate) and Figure 5.13c

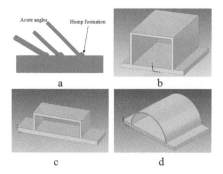

**FIGURE 5.13**
Fabrication of overhanging feature by the flat table (a) inclined wall by inclined torch, (b) horizontal wall with a horizontal torch, (c) horizontal wall with an inclined torch, and (d) semicircular box. (Adapted from Kazanas et al. 2012.)

(30° from the horizontal plane). The semicircular component where the inclination angle varies from 0° to 90° is also built from both ends as a quarter circle and joins at the middle by first forming the 60° *v*-grove as shown in Figure 5.13d (Kazanas et al. 2012).

Rotating table and vertical torch position (Panchagnula and Simhambhatla 2016, 2018) allows the most stable metal deposition as the weld pool remains in the horizontal plane. Based on the adapted inclination angle, truncated cone (single overhang), triangular duct (two overhangs), and semicircular duct (non-uniform overhang) can be deposited which are shown in Figure 5.14a, b, and c,

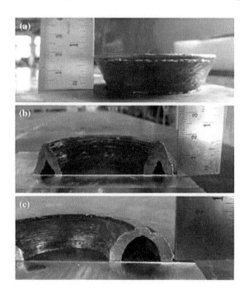

**FIGURE 5.14**
Overhanging features: (a) inverted truncated cone, (b) triangular duct, and (c) semicircular duct (Panchagnula and Simhambhatla 2016). (With the permission of Springer.)

respectively. The complex closed shapes are also possible using rotating table and vertical torch position approach, as shown in Figure 15.15a,b wherein full closing is restricted due to the accessibility of torch as shown in Figure 5.15c.

The rotating table and inclined torch allow fabricating complex feature components in a slightly easier manner. The object containing diverging, converging, and overhanging features were manufactured with a 6-axis robot and a 2-axis rotating table using GTAW WAAM (Bonaccorso et al. 2011). The propeller, which has an intricate revolved part, contains a curvature and overhang feature, as shown in Figure 5.16, is built with a laser-based direct

a                      b                      c

**FIGURE 5.15**
Fabrication of complex closing shape by rotating table and vertical torch; (a) closing cone, (b) closing sphere, and (c) accessibility issue in sphere closing (Panchagnula and Simhambhatla 2018). (With the kind permission of Elsevier.)

**FIGURE 5.16**
Fabrication of complex closing shape having a revolving feature by rotating table and vertical torch using LBDMD process (Ding et al. 2017). (With the permission of Elsevier.)

metal deposition (LBDMD) process. During the fabrication of the propeller, the 6-axis robot arm is coupled with 2-axis tilt and rotatory positioning base. The hybrid slicing method is established to map the overhanging structures of a revolved part to be at a planar base, which makes the traditional path planning strategies suitable for generating the tool path for the mapped structures (Ding et al. 2017).

## 5.4 Recent Developments in the Fabrication of Parts by WAM Process

### 5.4.1 Advance Power Sources

The development of power source technology played a significant role in the realization of WAM, particularly in the case of the arc-based deposition method. On comparing with other alloys, fabrication of aluminum alloy using the AM is difficult. This is mainly due to the porosity—a common issue faced with the AM. In order to overcome, an advancement in power source like CMT allows fabrication with porosity and spatter-free component (Cong et al. 2015). The CMT is highly efficient in heat management. The controlled dip transfer coupled with the mechanical movement of the wire (back and forth motion) allows maximum use of arc heat and elimination of spatters. Another method of minimizing the heat dissipation to the substrate is the use of double electrode GMAW without reducing the wire melting and the deposition rate is shown in Figure 5.17. A bypass GTAW torch is added to offer the second loop to supply the current to welding wire and decouple the base metal current from the melting current by introducing the bypass arc. This reduces the base metal heat input without reducing the wire melting and deposition rate. The current flow in the wire is the sum of

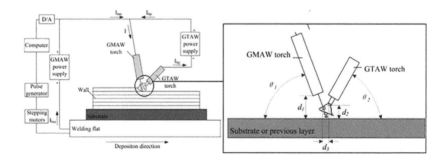

**FIGURE 5.17**
Schematic diagram of double-electrode GMAW-based AM system showing the relative position of two torches in zoom view (Yang et al. 2016). (With the kind permission of Elsevier.)

the base metal current ($I_m$) and bypass current ($I_b$). Without the bypass arc, the base metal current is equal to the melting current. The addition of the bypass arc reduces the base metal current. Thus, the width of the thin wall decreases with increasing bypass current, whereas wall height increases and height variation along the direction of deposition also decreases (Yang et al. 2016).

## 5.4.2 In-Situ Sensing and Adaptive Process Control

The geometric accuracy and defect-free fabrication along with high productivity are vital to match the requirement of the AM parts, which can be achieved by in-situ sensing during fabrication and adaptive process control. For example, the bead profile is very sensitive to the process parameters in WAAM wherein in-situ bead profile sensing by laser-based equipment can provide an adaptive control of the process. This adaptive control of WAAM process is executed by the advanced computer interface, as shown in Figure 5.18. The interface provides the facility such as bead modeling, slicing, path planning, weld setting, post-processing, and robot code generation (Ding et al. 2016a). The application of in-situ sensing and adaptive process control has the potential to regulate the WAM processes, which is expected to provide broader acceptability of WAM process in manufacturing industries.

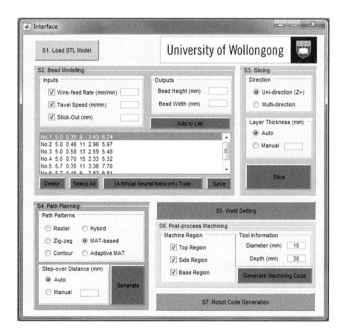

**FIGURE 5.18**
User-friendly interface of robotic WAAM system (Ding et al. 2016a). (With the kind permission of Elsevier.)

### 5.4.3 Hybrid WAM Processes

The WAM processes are hybridized with other fabrication operations in order to improve the efficacy. In milling-assisted WAM, the parts are milled after full deposition (Akula and Karunakaran 2006) or milled immediately after each layer deposition (Xiong et al. 2009). The latter cuts down the manufacturing efficiency due to time consummation in the shifting of the workpiece between the deposition and milling processes, however, helps in maintaining the accuracy. Milling-assisted WAM enhances the dimensional accuracy and surface quality, which creates a balance between the product qualities and fabrication time.

Commercially milling-assisted powder WAM is also available in the market. LUMEX Avance-25 and LUMEX Avance-60 are laser-based powder bed technology with milling capabilities developed by Matsuura Machinery Corporation (Matsuura Machinery, 2018). The machine tool manufacturer DMG developed LASERTEC 65 is another hybrid AM machine in which powder nozzle and laser are mounted on a 5-axis milling machine with a working area of workpieces up to ø 650 mm, 360 mm height and a maximum weight of 1,000 kg. It can generate parts for critical applications from precious and high-temperature alloys with the cost efficiency at a faster speed than the powder bed (Newman et al. 2015). The milling-assisted WAAM process is not yet commercially available. However, the welding gun can be mounted on the vertical milling center to realize milling-assisted WAAM. In milling-assisted WAAM, after deposition of each layer, the surface is facing milled that compensates the variation in thickness due to irregular weld surface. The complex geometry of the injection mold with core and cavity (Figure 5.19) is fabricated by milling-assisted WAAM (Karunakaran et al. 2010). The milling-assisted WAAM provides better economy and manufacturing time viability as compared to CNC machining (Song et al. 2005).

The hybridization of SLM and WAAM has also been reported. Fabrication of large and complex component by SLM and higher accuracy with WAAM are still to be achieved. The hybridization of SLM and WAAM can fabricate large and complex component economically. The parts fabricated by this process are decomposed into the complicated and simple part, which are generated by using SLM and WAAM, respectively. However, the inconsistency in mechanical properties is an unresolved issue—e.g., the tensile fractures occurring in WAAM zone (Shi et al. 2017).

### 5.4.4 Rolling-Assisted WAM

The top surface of the bead in WAAM is convex and non-smooth, which causes the flow of molten material and yields difference in theoretical and actual thickness resulting in a cumulative error in the built-up height direction. The in-situ rolling or intermediate rolling after each layer or layers makes the top surface almost smooth and flat, and alleviate the above-mentioned effect, which is shown in Figure 5.20. In-situ rolling controls

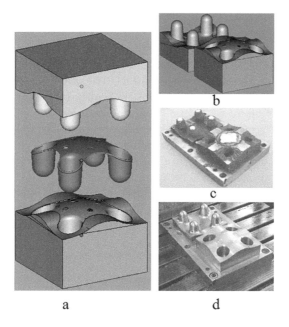

**FIGURE 5.19**
Fabrication of molding dies of massager components by milling-assisted WAAM (a) massager and it dies, (b) dies arrangement for assisted WAAM fabrication (c) near-net shape of the mold and die, and (d) finished parts of mold and die (Karunakaran et al. 2010). (With the kind permission of Elsevier.)

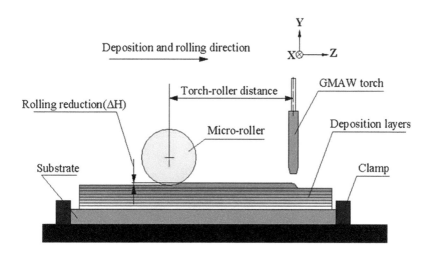

**FIGURE 5.20**
Schematic diagram of rolling on weld bead in rolling-assisted WAAM (Zhou et al. 2016). (With the kind permission of Springer.)

the deposition layer thickness (Zhang et al. 2013) and improves the deposition efficiency.

The position of the roller during in-situ rolling is crucial because too near a roller damages the bead; too far has under deformation and an appropriate roller distance produces the desired effect. The weld bead without rolling and with rolling is shown in Figure 5.21a and b, respectively (Zhou et al. 2016). The rolling load and roller size determines the optimum distance between the heat source and the roller. The optimum distance is in the range of 20–25 mm for micro roller during in-situ rolling.

Flat, profiled, pinch, inverted profile, and grooved roller are used for rolling-assisted fabrication of WAM component (Colegrove et al. 2017) which are shown in Figure 5.22. The "profiled" roller (i.e., similar to bead profile) is used to engage the convex top surface of the bead (Figure 5.22a). The rolling can be

**FIGURE 5.21**
Effect of the roller on weld bead in rolling-assisted WAAM (Zhou et al. 2016). (With the kind permission of Springer.)

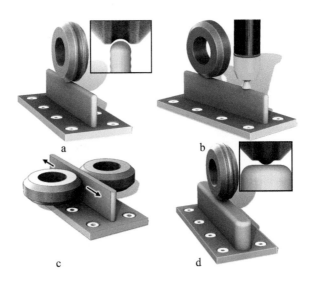

**FIGURE 5.22**
Various types of typical roller and rolling methods: (a) profile roller with vertical rolling, (b) in-situ rolling, (c) pinch rolling, and (d) inverted profiled roller for rolling thick sections and intersections (Colegrove et al. 2017). [With the kind permission of Elsevier (Open access), www.sciencedirect.com/science/article/pii/S1359646216305267, permission is not required.]

performed in-situ—followed by the welding torch (Figure 5.22b). The pinch rollers operate just ahead of a flat roller simultaneously (Figure 5.22c). The "slotted" roller prevents lateral distortion of the deposition during rolling (Figure 5.22d).

In addition to shape control, the benefit of rolling includes reduction of residual stresses and distortion, grain refinement, and elimination of big pore that enhance the mechanical properties.

Rolling is more effective in controlling residual stress reduction as a post-weld operation rather than in-situ rolling. Rolling at too high temperature overwhelms the effective strain of the material induced by rolling as it cools. The stresses in the WAM component without rolling reduce rapidly in a reasonably linear fashion, which is close to zero near the top of the wall. The slotted roller produces a maximum reduction in distortion and an increment in deposition efficiency (Colegrove et al. 2013). The rolling increases the layer width but decreases the layer height, distortion, and residual stress considerably with an increase of rolling load (Martina et al. 2016). In this section, the focus is given to the effect of rolling on residual stress and distortion.

## 5.5 Summary and Conclusion

WAM is a multidisciplinary process driven by the specific applications. Several versions of WAM equipment such as powder and wire-based AM are commercially available, and their numbers are continuously growing. Continuous improvement in automation, computational hardware, and software, arc power source (such as CMT), and higher knowledge in metallurgy and welding are the synergistic aspects for the rapid development of WAM. Defects in components such as lack of fusion, voids, porosity, poor surface finish, dross formation, distortion, and residual stresses are the primary concerns that need to be addressed to facilitate a broader commercial adaptation of WAM. The component of WAM should be as close as to 3D CAD model and proper planning of the deposition path can minimize the distortion to a diminishing value. The path planning depends not only on the geometry of parts but also on thermal management and on the ability of the process to change the bead dimension by the requirement. Fabrication of overhanging parts requires special attention to avoid the flow of the molten weld pool. In powder bed, the optimization of built-up orientation and support design reduce the distortion. The fabrication of overhanging features in wire-feed-based AM is slightly easier by using a rotating table and/or inclined torch. The top surface of the bead in case of wire-feed-based AM can be made flat by using the milling or rolling as an assisting element. The usage of assisted milling makes the surface quality of WAM parts to be an excellent grade of utility in actual application. The rolling-assisted manufacturing enhances the mechanical property. The broader market share of WAM will require a high level of control and standardization to achieve repeatability in component fabrication with consistent properties. However, certification and qualification are challenging tasks due to high cost and long-time consumption for extensive adoption of critical parts. Enhanced sensing process and an adaptive control system are very crucial to regulate the WAM processes for achieving high productivity without common defects. Ultimately, a comprehensive understanding and innovations are required for larger technological adaptation of the wire-based AM.

## References

Akula, S., and K. P. Karunakaran. "Hybrid adaptive layer manufacturing: An intelligent art of direct metal rapid tooling process." *Robotics and Computer-Integrated Manufacturing* 22, no. 2 (2006): 113–123.

A. S. T. M. Standard F2792. "Standard terminology for additive manufacturing technologies." ASTM F2792-10e1 (2012).

Bikas, H., P. Stavropoulos, and G. Chryssolouris. "Additive manufacturing methods and modelling approaches: A critical review." *The International Journal of Advanced Manufacturing Technology* 83, no. 1–4 (2016): 389–405.

Bonaccorso, F., L. Cantelli, and G. Muscato. "Arc welding control for shaped metal deposition process." *IFAC Proceedings Volumes* 44, no. 1 (2011): 11636–11641.

Calignano, F. "Design optimization of supports for overhanging structures in aluminum and titanium alloys by selective laser melting." *Materials & Design* 64 (2014): 203–213.

Chen, H., D. Gu, J. Xiong, and M. Xia. "Improving additive manufacturing processability of hard-to-process overhanging structure by selective laser melting." *Journal of Materials Processing Technology* 250 (2017): 99–108.

Cheng, B., and K. Chou. "Geometric consideration of support structures in part overhang fabrications by electron beam additive manufacturing." *Computer-Aided Design* 69 (2015): 102–111.

Chivel, Y., and I. Smurov. "Temperature monitoring and overhang layers problem." *Physics Procedia* 12 (2011): 691–696.

Colegrove, P. A., H. E. Coules, J. Fairman, F. Martina, T. Kashoob, H. Mamash, and L. D. Cozzolino. "Microstructure and residual stress improvement in wire and arc additively manufactured parts through high-pressure rolling." *Journal of Materials Processing Technology* 213, no. 10 (2013): 1782–1791.

Colegrove, P. A., J. Donoghue, F. Martina, J. Gu, P. Prangnell, and J. Hönnige. "Application of bulk deformation methods for microstructural and material property improvement and residual stress and distortion control in additively manufactured components." *Scripta Materialia* 135 (2017): 111–118.

Cong, B., J. Ding, and S. Williams. "Effect of arc mode in cold metal transfer process on porosity of additively manufactured Al-6.3% Cu alloy." *The International Journal of Advanced Manufacturing Technology* 76, no. 9–12 (2015): 1593–1606.

Ding, D., Z. Pan, D. Cuiuri, and H. Li. "Wire-feed additive manufacturing of metal components: Technologies, developments and future interests." *The International Journal of Advanced Manufacturing Technology* 81, no. 1–4 (2015a): 465–481.

Ding, D., Z. Pan, D. Cuiuri, and H. Li. "A multi-bead overlapping model for robotic wire and arc additive manufacturing (WAAM)." *Robotics and Computer-Integrated Manufacturing* 31 (2015b): 101–110.

Ding, D., C. Shen, Z. Pan, D. Cuiuri, H. Li, N. Larkin, and S. V. Duin. "Towards an automated robotic arc-welding-based additive manufacturing system from CAD to finished part." *Computer-Aided Design* 73 (2016a): 66–75.

Ding, D., Z. Pan, D. Cuiuri, H. Li, and N. Larkin. "Adaptive path planning for wire-feed additive manufacturing using medial axis transformation." *Journal of Cleaner Production* 133 (2016b): 942–952.

Ding, Y., R. Dwivedi, and R. Kovacevic. "Process planning for 8-axis robotized laser-based direct metal deposition system: A case on building revolved part." *Robotics and Computer-Integrated Manufacturing* 44 (2017): 67–76.

Dongdong, G. "Laser additive manufacturing (AM): Classification, processing philosophy, and metallurgical mechanisms." In G. Dongdong (ed.). *Laser Additive Manufacturing of High-Performance Materials*, 15–71. Berlin: Springer, 2015.

Gibson, I., D. W. Rosen, B. Stucker. Additive Manufacturing Technologies. New York: Springer, 2010.

Hascoet, J. Y., K. P. Karunakaran, and S. Marya. "Additive manufacturing viewed from material science: State of the art & fundamentals." *Materials Science Forum*, Vol. 783, 2347–2352. Zurich: Trans Tech Publications, 2014.

Hussein, A., L. Hao, C. Yan, R. Everson, and P. Young. "Advanced lattice support structures for metal additive manufacturing." *Journal of Materials Processing Technology* 213, no. 7 (2013): 1019–1026.

Jin, G. Q., W. D. Li, and L. Gao. "An adaptive process planning approach of rapid prototyping and manufacturing." *Robotics and Computer-Integrated Manufacturing* 29, no. 1 (2013): 23–38.

Karunakaran, K. P., S. Suryakumar, V. Pushpa, and S. Akula. "Low cost integration of additive manufacturing and subtractive process for hybrid layered manufacturing." *Robotics and Computer Integrated Manufacturing* 26, no. 5 (2010): 490–499.

Kazanas, P., P. Deherkar, P. Almeida, H. Lockett, and S. Williams. "Fabrication of geometrical features using wire and arc additive manufacture." *Proceedings of the Institution of Mechanical Engineers, Part B: Journal of Engineering Manufacture* 226, no. 6 (2012): 1042–1051.

Loughborough University, AMRG Group. The 7 categories of additive manufacturing. www.lboro.ac.uk/research/amrg/about/the7categoriesofadditivemanufacturing/powderbedfusion/ (accessed on 24 January 2018).

LUMEX Avance-60—CNC Milling—Matsuura Machinery. www.matsuura.co.uk/cnc-milling/lumex-avance-60/ (accessed on 24 January 2018).

Martina, F., M. J. Roy, B. A. Szost, S. Terzi, P. A. Colegrove, S. W. Williams, P. J. Withers, J. Meyer, and M. Hofmann. "Residual stress of as-deposited and rolled wire+arc additive manufacturing Ti–6Al–4V components." *Materials Science and Technology* 32, no. 14 (2016): 1439–1448.

Newman, S. T., Z. Zhu, V. Dhokia, and A. Shokrani. "Process planning for additive and subtractive manufacturing technologies." *CIRP Annals-Manufacturing Technology* 64, no. 1 (2015): 467–470.

Panchagnula, J. S., and S. Simhambhatla. "Feature based weld-deposition for additive manufacturing of complex shapes." *Journal of the Institution of Engineers (India): Series C* (2016):285–292.

Panchagnula, J. S., and S. Simhambhatla. "Manufacture of complex thin-walled metallic objects using weld-deposition based additive manufacturing." *Robotics and Computer-Integrated Manufacturing* 49 (2018): 194–203.

Rombouts, M., L. Froyen, A. V. Gusarov, E. H. Bentefour, and C. Glorieux. "Photopyroelectric measurement of thermal conductivity of metallic powders." *Journal of Applied Physics* 97, no. 2 (2005): 024905.

Shi, X., S. Ma, C. Liu, Q. Wu, J. Lu, Y. Liu, and W. Shi. "Selective laser melting-wire arc additive manufacturing hybrid fabrication of Ti-6Al-4V alloy: Microstructure and mechanical properties." *Materials Science and Engineering: A* 684 (2017): 196–204.

Song, Y. A., P. Sehyung, and S. W. Chae. "3D welding and milling: Part II—optimization of the 3D welding process using an experimental design approach." *International Journal of Machine Tools and Manufacture* 45, no. 9 (2005): 1063–1069.

Venuvinod, P. K., and W. Ma. *Rapid Prototyping: Laser-Based and Other Technologies.* Berlin: Springer, 2013.

Xiong, X., H. Zhang, and G. Wang. "Metal direct prototyping by using hybrid plasma deposition and milling." *Journal of Materials Processing Technology* 209, no. 1 (2009): 124–130.

Xiong, J., Y. Lei, H. Chen, and G. Zhang. "Fabrication of inclined thin-walled parts in multi-layer single-pass GMAW-based additive manufacturing with flat position deposition." *Journal of Materials Processing Technology* 240 (2017): 397–403.

Yang, D., C. He, and G. Zhang. "Forming characteristics of thin-wall steel parts by double electrode GMAW based additive manufacturing." *Journal of Materials Processing Technology* 227 (2016): 153–160.

Yilmaz, O., and A. A. Ugla. "Shaped metal deposition technique in additive manufacturing: A review." *Proceedings of the Institution of Mechanical Engineers, Part B: Journal of Engineering Manufacture* 230, no. 10 (2016): 1781–1798.

Zhang, H., X. Wang, G. Wang, and Y. Zhang. "Hybrid direct manufacturing method of metallic parts using deposition and micro continuous rolling." *Rapid Prototyping Journal* 19, no. 6 (2013): 387–394.

Zhong, Y., L. E. Rännar, S. Wikman, A. Koptyug, L. Liu, D. Cui, and Z. Shen. "Additive manufacturing of ITER first wall panel parts by two approaches: Selective laser melting and electron beam melting." *Fusion Engineering and Design* 116 (2017): 24–33.

Zhou, X., H. Zhang, G. Wang, X. Bai, Y. Fu, and J. Zhao. "Simulation of microstructure evolution during hybrid deposition and micro-rolling process." *Journal of Materials Science* 51, no. 14 (2016): 6735–6749.

# 6

## *Cares to Deal with Heat Input in Arc Welding: Applications and Modeling*

**Américo Scotti**

*University West*

*Federal University of Uberlandia*

### CONTENTS

6.1    Are Heat Input and Arc Efficiency Measurable Values?...................... 101
6.2    Heat Flux Inside a Plate during Welding ................................................ 104
6.3    Intrinsic Errors on Calorimetry When Heat Input Is Measured ........ 106
6.4    Gross Heat Input in Arc Welding: Meaning, Determination, and
       Application ................................................................................................. 110
       6.4.1    Determination of Intrinsic Error due to Weld Bead Length ....... 112
       6.4.2    Determination of Intrinsic Error due to Elapsed Time and
             Trajectory ...................................................................................... 114
       6.4.3    Determination of Intrinsic Error due to Test Plate
             Dimensions .................................................................................. 117
             6.4.3.1    Error due to Test Plate Thickness ................................. 117
             6.4.3.2    Error due to Test Plate Width ....................................... 119
             6.4.3.3    Error due to Test Plate Length ...................................... 122
6.5    Which Is the More Adequate Representation of Current When
       Heat Input into the Plate Is Concerned, Though Mean or RMS
       Values? ........................................................................................................ 125
6.6    Summary .................................................................................................... 135
References .............................................................................................................. 135

## 6.1 Are Heat Input and Arc Efficiency Measurable Values?

Heat input is an important phenomenon in the welding processes which calls attention to those involved with this technology. The precise understanding of this phenomenon is of utmost importance, to ensure that devised welding procedures are leading to quality welds. The solidification mode and secondary metallurgical transformations are affected by heat input. If the weld is performed with high heat input, the time that the metal stands above a temperature in which the grain growth kinetic is favored is increased. Accordingly, this characteristic increases the metal quench ability and the affected region will present less toughness, although hardness and ultimate tensile strength can reach opposite trends. A lower heat input, on the other hand, leads to fast cooling rates, which, depending on the material composition, can also quench the affected weld zones. Therefore, determination of heat input in welding in an accurate way is of major relevance.

In spite of innumerable attempts to further develop the subject, reinforced by the high technological status of welding, there are still some misinterpretations. One of them is a definition of the heat input and its connected basics (such as arc power or welding energy) and derivatives (such as melting and arc welding efficiency). These terms are widely applied in studies to correlate welding parameters to metallurgical transformations. Although the American Welding Society (2001) clearly defines "heat input" as energy delivered to the workpiece, often the term heat input is replaced by "welding energy" and vice versa. For instance, several authors use an equation of welding energy (to be demonstrated ahead) and name the outcome as heat input while studying the influence of heat input on microstructure and mechanical properties of welded joints. One can find such other definitions for heat input as "power input" (Lu and Kou, 1989), "true heat input" (Joseph, 2001), and "energy input" (Zeemann, 2003; Gery et al., 2005; Haelsig et al., 2011).

In the same perceptive, Quintino et al. (2013) warns, nevertheless, that a comprehensive knowledge of the other parameter effect on the heat input into the plate, such as plate thickness and the type of pass to be performed (full penetration, root pass or filler pass), is also required if a less-conservative welding procedure is aimed. According to them, welding engineers should also consider the effect of heat losses through the back side of the weld by radiation for full penetration welding, when developing welding procedure specifications (WPSs) based on partial penetration welds. As seen in Figure 6.1, the same arc energy produced different welds

**FIGURE 6.1**
Transversal sections of plain carbon steel short-circuiting MIG/MAG welding (150 A, 20 V, 34 cm/min, Ar + 25CO$_2$, 1, 2 mm AWS ER70S-6 wire): (a) full penetration welding and (b) partial penetration welding (Quintino et al., 2013). (With kind permission from Elsevier.)

in a material with mismatching thickness; actual heat input and heat flux were likely the cause.

A remarkable number of research projects and publications have been dedicated to the measurement of heat input from arc welding processes. The main reason would be the need for a more accurate parameter to be used in process applications and modeling. A lot of devices (e.g., Nasiri et al., 2014) and measurement methods (e.g., Wong and Shih, 2013) have been proposed over the years. As welding process modeling is progressively reaching a stage of practical usage, the importance of a precise determination of heat input and its correlated arc efficiency is getting more evident. It is important for those who are modeling temperature-dependent phenomena inside the base metal to know a precisely defined value of the heat input. An underestimated value of this heat input, as when heat losses from the plate surfaces cannot be avoided during the calorimetric measurements, may lead to imprecise predictions. The simulation models, indeed, calculate the heat losses from the plate surfaces, implying that part of the losses is duplicated if the heat input parameter is marred by losses.

Also arc efficiency became the most popular outcome of heat input measurements, the great discrepancies among the results of calorimetric measurements (see Stenbacka, 2013) and the lack of awareness of the heat losses in the measurement outcomes have left many researchers skeptical about the existence of reliable values of heat input or arc efficiency to be applied in their models. In practice, this parameter is usually tuned so that the simulation results reproduce weld geometries observed experimentally. A precise value of the heat input would allow eliminating this specific tuning and make the models more predictive.

Liskevych et al. (2013) claimed and provided evidence that intrinsic errors (mainly related to test plate geometry) are due to calorimetric measurement procedures. Their findings are in good agreement with those of Stenbacka (2013), for whom uncontrolled heat losses occur during welding (from the sample surfaces to the environment or fixtures), as well as during the transportation of the sample to the calorimeter (when this is necessary). This source of errors might be the reason for the above-mentioned discrepancies among the results of calorimetric measurements. This explains the existence of reported attempts to improve calorimeters by minimizing heat losses from the plate surfaces. Egerland and Colegrove (2011), for instance, detailed how they reduced further heat losses to the surroundings using an insulated box calorimeter. Haelsig and Mayr (2013) used in their studies a modified water calorimeter, in which the plate is inclined and the water level in the vessel is increased constantly and in phase with the heat source movement. They claim that this setup guarantees that the introduced heat is directly transferred into the calorimetric medium, and no insulating air film reduces the heat transfer. However, they recognize that energy losses to the surrounding atmosphere during and after welding are still sources of error. Pépe et al. (2011) consider that the noticeable spread in calorimetric results arises from both systematic and random errors. The systematic error, in special, can underestimate the actual heat input.

## 6.2 Heat Flux Inside a Plate during Welding

As can be seen in the introductory section, there are different ways for dealing with heat input, from terminology to measurement points of view. Despite the wide usage of the heat input parameter (WPS, modeling, welding metallurgy studies, etc.), there is no consensus in the current literature about how heat flows inside a plate during the welding. Considering the several means of heat transfers in welding system, Figure 6.2 was elaborated (Scotti et al., 2012) and updated (Hurtig et al., 2016) with the ambition of showing qualitatively that just a part of the arc energy is transferred into the plate (yet recognized in quantitative analysis as the most significant). Arc energy is transformed into heat at the arc-plate coupling (electrical current dependent). However, arc energy is also responsible for melting the feeding wire, which

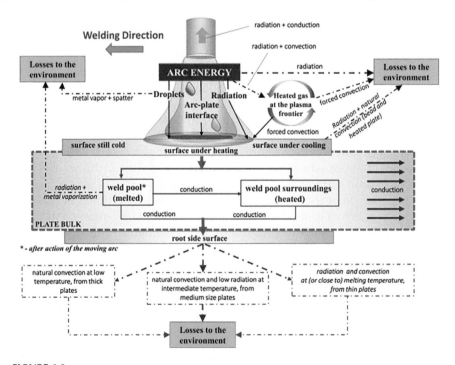

**FIGURE 6.2**

Schematic representation of the heat input and output in a plate during moving arc welding (full lines: input; dotted lines: output; italic text: output that does not influence near bead cooling rates) (Hurtig et al., 2016). (With kind permission from Taylor & Francis.)

heat content for the wire melting (mainly sensible and fusion latent heats) is transferred to the weld pool (through droplets). The arc also irradiates directly heat to the pool. Complementing, the heat to transform the matter into plasma and that warms the shielding gas beyond the plasm frontier is also transferred to the plate by forced convection (plasma jet).

However, a parcel of the energy of the arc is lost to the wire, cooled contact tip (by radiation and conduction) and also to the environment (by radiation). Since the Arc jet deviates sidewise when hitting the plate/pool, a part of internal heat energy of plasma and shielding gas at the plasma frontier is lost to environment. Part of the internal heat of the plasma and of the warm shielding gas at the plasma frontier goes to the environment since the arc jet deviates sidewise when hitting the pool/plate. Metal vapor and spattering are another source of heat that dissipates to the environment before being transferred to the plate. There is still an aspect of the heat losses during arc welding that is not widely commented in the analysis of arc efficiency (traditionally expressed as the ratio of heat input and the arc energy), i.e., the heat that after having entered the plate goes also to the environment. The weld pool loses some amount of heat to the environment by radiation and metal vaporization after the arc is moved away from the pool (during solidification), rather than having the whole heat conducted into the plate bulk. Even the heated plate, on the upper side and on the back side, undergoes heat transfer to the environment. Depending on the plate thickness and the diffusivity coefficient, the losses of heat at the plate back side can take place by the following three means: (1) natural convection losses with no importance at low temperature if the plate is thick enough to not have the surface significantly heated (for a given arc energy), (2) radiation and convection losses at high temperature from the root bead pool and convection losses (solidified root bead) if the plate is thin enough to have the opposite surface undergoing melting temperature, and (3) natural convection at median temperature if the plate thickness is intermediate. Normally, calorimetric heat input determinations are not able to sort the different heat parcels into the material.

Last but not least, there is another issue not clearly discussed by welding research community, which is related to the role of the heat input on the material. From the metallurgical transformation point of view, the important is the resultant heating and cooling rates that the heat input will impose on the material. Thus, heat which is lost to the environment even after entering the plate will not highly influence the cooling rate (the effect would be on delaying slightly the heat conduction), because it is lost before diffusing into the material. On the other hand, from the bead formation and/or thermal stress generation, the heat that momentously stayed in the plate may play an important role. The heat distribution along the plate thickness is another issue. The heat fluxes at a different position of the weld bead or heat affect zone of the same weld are certainly different. In fact, the actual heat and its role in the welding behavior are not a straightforward knowledge, as the welding personnel could wish or hope to be.

## 6.3 Intrinsic Errors on Calorimetry When Heat Input Is Measured

In relation to the measurement of the thermal efficiency factor (consequently, the heat input), a review of the technical literature shows that some debate has taken place concerning the most appropriate method to obtain a correct measurement. Bosworth (1991) and Essers and Walter (1981) used classical calorimeters (water in an insulated vessel), where the monitoring of the water temperature was the main parameter to determine the heat involved in the welded parts immersed in the flask. However, one of the most used techniques is the water calorimeter, as described by Lu and Kou (1989), which allows a continuous cooling with continuous water flow through the root side of the weld (the water temperature during and after welding is monitored and the heat absorbed by the running water). A similar principle was used by Cantin and Francis (2005), who used as heat absorber a block of electrical conductor grade aluminum, over which it was attached the plate to be welded (all inside an isolated box). Pépe et al. (2011) improved Cantin and Francis's calorimeter by using a movable lid to close the upper surface of the isolated box just behind the arc movement.

Another class of calorimeter used to measure heat efficiency was described and applied by Giedt et al. (1989), Fuerschbach and Knorovsky (1991), and DuPont and Marder (1995). This calorimeter consists of a special box (in which the hot workpiece is placed inside) and works based on the Seebeck principle (voltage is produced proportionally to the heat flux through the calorimeter walls, i.e., gradient layer principle). Soderstrom et al. (2011) used another interesting calorimeter for the measurement of droplet heat content. Haelsig et al. (2011) presented another approach of water calorimeter, in which the metal plate is positioned at a certain angle in the calorimetric vessel. Concurrently to the movement of the welding torch, the water level in the calorimetric vessel is constantly increased. With this design, they claim that the heat from the heated portion of the metal would be totally transferred to the circulating water. Nonetheless, heat losses that occur during welding and/or in the elapsed time between finishing the weld and the measurement seem to be a setback shared by most of, if not all, calorimetric techniques.

Cryogenic calorimeters have also been used by several researchers in welding (Watkins, 1989; Hsu and Soltis, 2002; Joseph et al., 2003; Pépe et al., 2011) with good results and with the advantage of reducing the time of the experiments. The plate is rapidly inserted into a Dewar containing liquid nitrogen after welding and the amount of liquid boiled off is measured. But, even though a reasonably accurate procedure has been stated for this method, some intrinsic problems exist in the methodology. During the calorimetric measurement, the lack of attention to its limitations leads to controversial results. In a study by Bosworth (1991), using a classical water calorimeter was impaired by considerably long elapsed time (15 s) between the end of the welding and the start of

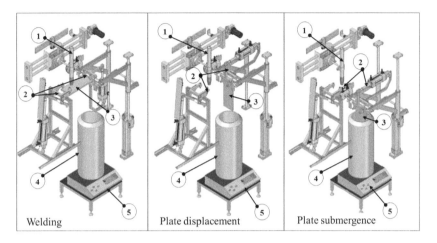

**FIGURE 6.3**
Schematic representation of the automated cryogenic calorimeter: 1—torch, 2—pneumatic system, 3—plate, 4—calorimeter flask, and 5—scale (Arévalo, 2011).

measurement. Lately, Pépe et al. (2011), using the advanced cryogenic calorimeter, also have shown that both the delay time between completing welding and inserting the specimen into the liquid nitrogen and the weld bead length influences the measured quantity. They indicate that the heat input reduces with increasing delay time or bead length (a drop of approximately 10% in the heat efficiency from delays changing from 5 to 100 s and from welds with 5 and 25 s of duration). The reduction in efficiency is caused by conduction from the sample into the jig, as well as convection from the sample after and during welding. These authors claim that for calculating the actual process efficiency it is necessary to subtract the errors due to both the welding and delay times. Arévalo and Vilarinho (2012) demonstrated a full-automated cryogenic calorimeter, as illustrated in Figure 6.3. They claimed that the random error becomes insignificant by using this device.

Pépe et al. (2011) also showed that different parameters (consequently, different welding energies) change the heat efficiency of a process. A reduction in efficiency was met for increasing welding energy, in agreement with Bosworth's (1991) results. One can suppose that if Pépe et al. used the same procedure and test plate sizes, a higher welding energy could represent bigger losses to the environment before digging the plate into the cryogenic vessel. DuPont and Marder (1995), using the Seebeck calorimeter, found, on the other hand, that the arc efficiency did not vary significantly with a process over the range of currents investigated. Pépe et al. further demonstrated that welding in a groove increased the process efficiency for around 5% since much of the radiation heat losses were absorbed by the side walls.

Other sources of errors are supposed to exist even in the sophisticated water calorimeter presented by Haelsig et al. (2012). The authors affirmed that since the water level in the calorimetric vessel was constantly increased, the heat loss occurring from the plate to the environment was avoided. However, even though it can be lower than in other calorimeters, there is still a distance between the plate at the location of heat input source (welding arc) and the water and not all heat is directly transferred into the calorimetric environment. Pépe et al. (2011) presented an unsuitable utilization of thermal properties of the materials to be welded as a source of errors even in an insulated box. A perfect isolation is not possible even in this type of calorimeter as the gradual closing of top cover inducts losses in the front of the welding torch as well as the losses to the surroundings. These types of errors are termed as intrinsic errors. These errors should not be confused with accuracy, which is related to random and systematic errors. A non-calibrated measurement system (source of systematic errors), or low accuracy instrumental (source of random errors) or high noises in the signal (source of systematic and random errors), or not precise settings of sensors and adjustment of the arc length (source of random errors), can lead to significant deviations from the true value. The intrinsic errors and the other random errors would be added to these errors.

The repeatability results and from calorimetric tests are not usually determined and provided in the open literature. However, in one of the studies by Pépe et al. (2011), 8% error was reported in replications while measuring cryogenic calorimetry. In addition to this, Stenbacka (2013) reported that up to 12% error in the measurements related to thermal efficiency can be accepted and defined as a fairly good execution. Sikström (2015) used a water-cooled stationary anode calorimeter as a case study to determine errors in the measurement process of arc efficiency. He focused his study on instrumental errors, operator influence (manual operations that introduce operator bias in the measurement process, such as arc length setting) and welding leads voltage drop (measuring the potential between the terminals of the welding power source instead of measuring the potential between the electrode contact tube and the workpiece). He showed that the operator bias is negligible in the estimate, but on the contrary, the consequence of neglecting welding lead to voltage drop results in a significant underestimation of the arc efficiency. His instrumental measurement error was estimated as ±0.03.

With the same objective, Cantin and Francis (2005) calculated the potential source of errors from current measurements as 1%. For them, the most significant sources of errors appear in the measurement of the voltage drop (the losses sum is up to 0.5 V). There was an uncertainty of approximately 0.2 mm in setting the arc length (leading to corresponding variations in an arc voltage of approximately 0.1 V with an argon shielding gas, and approximately 0.4 V with helium). In addition, these authors pointed out errors in estimating the temperature rise of the aluminum workpiece (the uncertainty in estimating the rise in enthalpy of the workpiece was approximately 3%). In turn, Haelsig and Mayr (2013) showed that the system that they used presents ±3% of the standard deviation. One percent of the measurement error is resulting from the range of accuracy of the thermocouples. Slightly more than 1% of the measurement error is introduced by the measurement of current and voltage. The measurement of the mass flow rate of the water contributes ±0.5% of uncertainty to the measurement.

Hurtig et al. (2016) applied a sensibility analysis to the most important measuring parameters of a water-cooled stationary anode calorimeter (voltage, current, water flow rate, water temperature difference at different water flow rates, and water specific heat capacity). Factors that could cause intrinsic error were not varied. They concluded that the test is very sensitive to the low accuracy of voltage, current, water flow rate, and temperature difference measurements (the error effect on arc efficiency calculation reached values as high as ±10%), although not significantly sensitive to the water specific heat capacity. However, voltage does not depend only on the measurement device (own accuracy), but also on the setting of the electrode tip to plate distance and the position of the voltage measurement.

The use of numerical simulation methods could present an advantage over calorimetric methods, based on their capacity to predict heat losses to the environment considering convection, conduction, and radiation means. However, these methods, even the latest ones, are subjected to other kinds of intrinsic errors. Carvalho et al. (2006) have evaluated the accuracy of their results and confirm a degree of sensitivity of their numerical models to variations of input parameters, such as thermal properties and variations of heat transfer coefficients with high temperature. In addition, as cited by Cângani et al. (2010), despite numerous pieces of work dedicated to the thermal analyses of welding processes, just a few were carried out considering phase transformations (which are a typical welding phenomenon). Several times, radiation losses (from both the arc and the welded metal) were not considered, as seen in the work by Lima e Silva et al. (2003). The inverse problem techniques that employ modern temperature measuring approaches (as thermal camera, used by Bardin et al., 2005; Nowacki and Wypych, 2010; or other optic systems), which can replace the traditional thermocouple method, are not reliable enough, once precise settings of the emissivity of the tested materials are needed, which are dependent on the environmental conditions. Furthermore, these methods have considerably high costs.

In summary, heat input and respective thermal efficiency values are always strongly dependent on the conditions in which they were determined (base material, sample dimensions, welding parameters, environmental conditions, etc.). It does not mean that calorimetry or numerical methods should not be used any longer. The important is to use them conscious of the measurement errors (intrinsic, random, and systematic errors). That is the reason for this author to believe that the best wording to define the results from calorimetry test would be "absorbed heat" (the heat quantity measured) rather than heat input, because there is no evidence that all the measured heat will influence the metallurgical transformations.

If the heat leaves the plate by the surfaces (edges, top, and root sides) before diffusing into the plate, this quantity should not be taken into account. For instance, with a long bead long during the measurements, the heat that is lost to the atmosphere during the welding through the surface, before digging the plate in the liquid nitrogen, can be significant, but it is not accounted as heat input. On the other hand, part of the heat that would go out by the root surface before diffusing is counted as heat inside the plate in the test. Thus, the absorbed heat measured in the test will be dependent on the bead length and plate dimension (thickness, width, and length) and shape. But it is not sure that absorbed heat would always represent the consequent cooling rates, which is of major importance to control the metallurgical transformations, as one would expect from the words heat input. The aspects referred above lead to the conclusion that the calorimetric tests can be used only for the comparative purpose, as long as all test parameters and welding conditions are the same.

## 6.4 Gross Heat Input in Arc Welding: Meaning, Determination, and Application

Considering that comprehensive models on heat transfer in welded plates take into account plate dimensions and that the amount of heat losses from the surfaces is estimated, the input values of the heat delivered to the plate must be free from this surface heat loss. One must remember that heat input is calculated as the product of the arc energy and a thermal efficiency factor, the latter obtained from calorimetry. However, as intrinsic errors, surface heat losses are partially accounted as absorbed heat in calorimetric measurements. Therefore, if the heat input values used in thermal-related welding models are not measured free of the heat losses through the plate surfaces, the thermal losses will be repeated (the calculated losses would be calculated assuming that no heat loss had happened in the thermal efficiency factor). Accordingly, if this fact is not observed, the target parameter from a model (cooling rates, for instance) may turn imprecise microstructures or bead and HAZ geometries, with consequences on the estimation of final microstructures/mechanical properties or deformations/residual stresses. Sikström et al. (2013) demonstrated that the measured cooling times of weld metal deviated expressively from those projected by calculations if a single value for heat input is used to all simulations, regardless of the welding conditions. Quintino et al. (2013), studying full penetration welding, used different heat efficiency factors rather than a generic one to show the importance of considering the role played by the heat losses by radiation through the weld backside.

Therefore, a new terminology is introduced in this section, i.e., "gross heat input." Gross heat input represents the amount of heat that is delivered on the test plate surface, but before happening any heat conduction inside the plate or loss of heat directly to the environment through the surfaces. This quantity could be termed as "net heat input," but to avoid confusion with a longer existing term meaning a difference between in and out heat, the first term was adopted. The common terminology "total heat input" was also avoided on purpose. It could indeed be misleading, since it is indifferently used to mean the heat that is input, for example, to the torch or to the workpiece altogether. Gross heat input is not measured directly by calorimeters, yet it can be indirectly determined by this means, a seen below.

The fact that intrinsic errors are still kept in measurements even when automated cryogenic calorimeter is applied was shown by Liskevych et al. (2013). However, these authors showed an approach to determine and eliminate these errors in the heat input quantification. Taking a cryogenic calorimeter as a case study (Figure 6.3), a well-grounded list of intrinsic errors was identified as follows (to remove these errors from calorimeter measurements is a purpose):

a. *Error due to operator influence.* The test plate after being welded needs to be moved from the plate holder to the calorimeter flask. During a manual transference of the test plate, not only the elapsed time but also the trajectory will be unlike in each measurement. Consequently, air convection will cool down the workpiece differently each time, which will result in a random error, which can be mitigated by automation of the operation. However, the sources for this error lead additively to a systematic error, since time and trajectory are not null and heat loss will take place.

b. *Error due to weld bead length.* If a too long bead is welded, significant heat losses, occurring through the test plate surfaces and the top surface of the bead, will take place before ongoing calorimetric measurement. Thus, long welds will always lead to lower values of absorbed heat per unit of bead length. On the other hand, as thermal equilibrium may not be reached if a really short bead is carried out, this condition does not characterize a representative welding process. This is typically an intrinsic systematic error of measurement.

c. *Error due to test plate dimensions.* Test plate thicknesses, widths, and lengths, irrespective of the bead length, will control the heat losses through the top, bottom, and side surfaces of the plate before ongoing calorimetric measurement. The adoption of a same dimension for the test plates would offer a feasible solution (standardized error). However, this error would be different according to the welding energy, unless the dimensions are too large to be considered thermally infinite for any welding parameters (unpractical for calorimeters). This error of measurement can also be labeled as an intrinsic systematic measurement.

Thus, the methodology, as proposed by Liskevych et al. (2013), is an approach for indirect determination of gross heat input values with a cryogenic (LN$_2$)

calorimeter, based on the absorbed heat per unit of bead length measurements and on the identification and removal of the intrinsic errors (systematic and random) of measurement. As already briefly mentioned, the principle of this liquid nitrogen calorimetric test is to quantify the heat absorbed by a plate during welding by plunging the welded plate into a Dewar flask full of a cryogenic liquid. The total absorbed heat by the test plate afterward is quantified by knowing the evaporation latent heat of the nitrogen, the plate mass, and the difference between the mass of the Dewar flask before and after welding. The absorbed heat by the welded plate (heat carried by the droplets, heat generated at the connection arc-plate and heat transferred from the plasma-gas enthalpy) is, then, determined by knowing the internal heating of the plate with the bead at room temperature (the $LN_2$ mass difference, before and after the specimen is dropped into the container).

Fundamentally, the actual gross heat input as proposed by Liskevych et al. (2013) is determined, for a given welding condition (process and parameters), by measuring the absorbed heats per unit of bead length, extrapolating it to null bead length and expurgating the intrinsic errors. To apply this method, several experimental steps must be followed. To make it clearer, a case study (Liskevych and Scotti, 2015) will be used to illustrate this procedure. The case study is related to a short-circuit MIG/MAG welding, with monitored parameters (mean values) of 150 A, 20 V, shielded by Ar + 25% $CO_2$ and using a 1.2-mm AWS ER70S-6 wire. For the gross heat input determination, a liquid nitrogen calorimeter and specimens of 200 mm × 100 mm × 6.3 mm made of plain carbon steel were used, with a fall distance (trajectory between the plate under welding and the Dewar flask, as illustrated in Figure 6.5) of 420 mm.

### 6.4.1 Determination of Intrinsic Error due to Weld Bead Length

To carry out several tests on the same plate thickness (the thicker the better), varying the bead lengths was assumed to be the solution for extrapolating the absorbed heat to null bead length. The author's experience suggests bead lengths from 5 to 100 mm, with steps progressively wider to better represent the trends and at the same time to mitigate the number of experiments. Consequently, the intrinsic error due to test plate thickness over the gross heat input value is minimized. In this case study, beads were deposited with lengths ranging from 5 to 112 mm. The results of the calorimetric measurements are given in Figure 6.4a. As can be seen from the figure, despite using the same welding energy, the absorbed heat per unit of weld length was lower for longer bead lengths. Thus, the intrinsic error of the weld bead length led to significant differences in the absorbed heat measurements.

Based on the curve trend in Figure 6.4a, it would be reasonable to assume that the influence of bead length on the value of the absorbed heat per unit of bead length would become insignificant when the time of welding is tending to 0. This quantity of heat from a welding time tending to 0 s was here denoted as gross heat input. Thus, gross heat input can be determined

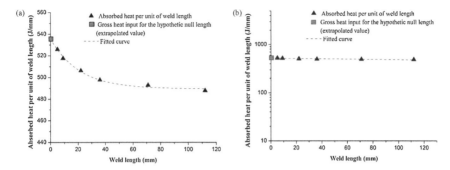

**FIGURE 6.4**
Representation of the extrapolation of the absorbed heat per unit of length from welds at different lengths, reaching the gross heat input (when weld length is tending to 0): (a) the fitted curve showing a natural exponential "e" behavior and (b) the same curve after a log-linearization (Liskevych and Scotti, 2015). (With kind permission from Elsevier.)

analytically by extrapolating the fitted curve to 0. As the function was a downward-sloping exponential, a linearization was applied in the Y-axis, as represented in Figure 6.4b. In the case of a natural exponential "e" function, the log-linearization converts the nonlinear equation into an equation that is linear in terms of log-deviations from their steady-state values. This approach of extrapolating the results for hypothetical beads with no lengths indicates the capacity of a process to input heat into the plate.

The extrapolated gross heat input values and statistical data are presented in Table 6.1. The choice of the curve equations was based on the arbitrary best fitting to the data. The statistical $R^2$ parameter is suitable for this task. As evident, bead length was the only input (independent variable) variable, while the absorbed heat was the only estimator (dependent variable). Gross heat input obtained reached a value of 534 ± 1 J/mm, applying the best-fit equations either before or after linearization. It can be concluded that the sensitivity to small variance is very low in cases in which the high-gradient part of a fitted curve is outlined by actual data and not only by extrapolated data. Thus, the extrapolation can be done by either using or not using linearization, leading to the same result, a procedure not possible in other cases, as it is going to be demonstrated ahead. For this reason, linearization was a standardized option adopted in this work (for clarification sake, however, extrapolations from both exponential and linearized equations will be shown along this case study), even at the cost of possible reduction of the data fitting correlation parameter ($R^2$). It is important to mention that this value for gross heat input is valid only for the presented measurement conditions and other intrinsic errors, such as elapsed time between the end of the welding and the start of the measurement, trajectory error and plate dimensions errors (length, thickness, and width) are not encompassed.

At this point, it is essential to justify using statistics, that the replications are not needed in the above and further experiments. According to

**TABLE 6.1**

Extrapolated Net Heat Input Values and Statistics, Related to Figure 6.4 (Mean Welding Energy of 621 J/mm, 420 mm of Fall Distance and Plate Dimensions of 200 mm × 100 mm × 6.3 mm)

| The Net Heat Input (J/mm) | Fitted Curve Equation | ANOVA | | | |
|---|---|---|---|---|---|
| | | DF | $R^2$ | F | StE |
| 534.0 | $y = 489.1 + 44.9e^{-x/22.0}$ | 3 | 0.986 | $3.0 \times 10^{-8}$ | 1.6 |
| 534.3 | $y = 534.3 - 0.4x$ | 5 | 0.905 | $1.7 \times 10^{-8}$ | 0.9 |

*Note:* DF, degree of freedom of the regression; $R^2$, data fitting correlation; $F \leq$ means that the estimator is statistically significant at 95% of confidence level; StE, predicted standard error of the estimator.

Liskevych et al. (2013), when there are a small number of factors (input variables), and higher number of factor levels (6 levels, i.e., 5–112 mm), the degree of freedom of the experiment is high enough. A consistent significance of the tendency can be reached if the hidden variance is low. In an analysis of variance, an *F*-test is often used to determine whether any group of trials differs significantly from an expected value. If the calculated ratio is less than the table value (Prob, for a significance level of, for instance, 95%), the null hypothesis that the variance is not significantly different is accepted. It means that not only does the equation fit the data (measured by $R^2$), but also the results do not differ from what was expected (Prob > *F* must be lower than 0.05).

However, the above-mentioned results are related to data in which the intrinsic errors due to either the test plate dimensions (width and length) or the elapsed time-trajectory have not been assessed.

### 6.4.2 Determination of Intrinsic Error due to Elapsed Time and Trajectory

As shown in the previous section, elapsed time and trajectory are random errors prone to be mitigating by maintaining the same elapsed time and trajectory during the test plates travel to the calorimeter flask. Nevertheless, as to make time and trajectory null (eliminating the systematic error of the quantities) in practice is unlikely, the proposal was to determine the weight of their effect on the absorbed heat per unit of bead length for different elapsed time and trajectory conditions and to extrapolate the result to a condition of null elapsed time and trajectory. This can be realized by having welds carried out with the same parameters on a plate of the same dimensions and the calorimetric measurements taken with different fall distances (Figure 6.5). For the case study, the fall distances were 550, 445, 420, and 395 mm (the liquid nitrogen level was always kept at the same level of 200 mm from the flask top). For each fall distance, the test was replicated using different bead lengths (so that extrapolation to bead length = 0 could be applied).

**FIGURE 6.5**
Schematic representation of the cryogenic calorimeter, with emphasis on the fixed trajectory of test plate under transference and the definition of the fall distances (Liskevych and Scotti, 2015). (With kind permission from Elsevier.)

The curves of absorbed heat per unit of weld length from different fall distances (already extrapolated to weld length tending to 0, with and without linearization) are shown in Figure 6.6. As can be seen in the figure, longer fall distances (i.e., the longer the time between the end of welding and the start of the measurement for a fixed trajectory) lead to less measured absorbed heat by the cryogenic calorimeter. This happens because heat loss arises when the welding arc is turned off, although the test plate has not yet arrived at the calorimeter flask. Furthermore, during the falling time, heat loss occurs more rapidly (compared to a stationary state of the test plate) due to the increase in the relative airflow speed (convection heat loss).

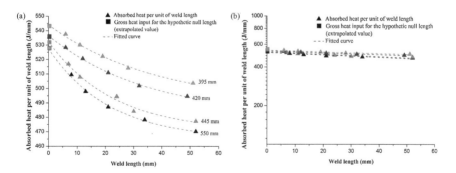

**FIGURE 6.6**
Heat absorbed by the test plate from welds at different lengths for various test plate fall distances and respective curve fittings: (a) the fitted curves showing a natural exponential "e" behavior and (b) the same curves after a log-linearization (Liskevych and Scotti, 2015). (With kind permission from Elsevier.)

It is evident from Figure 6.6 that the extrapolated values of absorbed heat curves to a weld length equal to 0, i.e., the determined gross heat input values, are different for each fall distance. A progressive reduction of the gross heat input values can be deducted as fall distances become longer. One explanation for this trend would be that a fall distance equal to 0 would lead to the highest gross heat input since this intrinsic error would also be eliminated. Consequently, another extrapolation must be applied over the fitted curve composed by the gross heat inputs from each fall distance to remove the error carried by elapsed times and trajectories. With this approach, the hypothetical condition in which calorimetric measurement would start exactly at the same moment that welding is finished (i.e., fall distance = 0). Figure 6.7 illustrates this reasoning, where, due to the nonexistence of actual data points close to the ordinate axis, a data linearization was applied.

Table 6.2 presents the best-fit equations (the first row with the natural exponential "e" behavior and the second row after log-linearization) and resultant

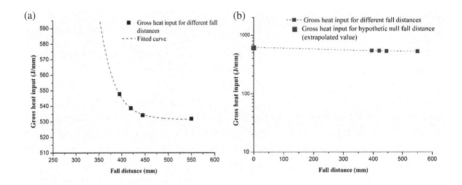

**FIGURE 6.7**
Extrapolation of the gross heat input values from different test plate fall distances to the fall distance equal 0 (gross heat input free from bead length and elapsed time-trajectory errors): (a) the fitted curve showing a natural exponential "e" behavior and (b) the same curve after a log-linearization (Liskevych and Scotti, 2015). (With kind permission from Elsevier.)

**TABLE 6.2**

Extrapolated Gross Heat Input Values and Statistics, Related to Figure 6.7 (Mean Welding Energy of 621 J/mm and Plate Dimensions of 200 mm × 100 mm × 7.95 mm)

| Gross Heat Input (J/mm) | Fitted Curve Equation | ANOVA | | | |
|---|---|---|---|---|---|
| | | DF | $R^2$ | F | StE |
| $1.6 \times 10^7$ | $y = 531.5 + 5.3 \cdot 0.7^x$ | 1 | 0.973 | $3.7 \times 10^{-6}$ | 0.4 |
| 573.2 | $y = 573.2 - 0.085x$ | 3 | 0.937 | $2.6 \times 10^{-7}$ | 4.1 |

*Note:* DF, degree of freedom of the regression; $R^2$, data fitting correlation; $F < 0.05$ means that the estimator is statistically significant at 95% of confidence level; StE, predicted standard error of the estimator.

predicted gross heat inputs at a fall distance equal to 0 reached by following the above-mentioned procedure. As evident from Table 6.2 and Figure 6.7, the sensitivity to small variance became too high and the results are subject to a high difference between predicted and actual values if linearization is not applied. Thus, only the equation from linearization is taken into account in this case. It is important to point out that, according to this methodology, the intrinsic error due to elapsed time and trajectory is eliminated, from a statistical point of view, along with the error due to the bead lengths.

### 6.4.3 Determination of Intrinsic Error due to Test Plate Dimensions

The effect of the test plate dimensions on the absorbed heat measurements and, consequently, on the gross heat input determination is also related to losses of heat within the time from welding start and the calorimetric measurement end. The use of a test plate with infinite dimensions (no heat losses through the sides and back surfaces before the start moment of the calorimetric measurement) would be an ideal procedure. Nonetheless, plate dimensions are constrained by the calorimeter, which presents limitations concerning test plate size and weight. The approach applied in this case study to determine the intrinsic errors due to test plate dimensions was grouped according to the source of error caused by each plate dimension, as follows.

#### *6.4.3.1 Error due to Test Plate Thickness*

The dimensions of the test plates (200 mm × 100 mm) for this case study were the same for all tests so that the influence of the plate thickness on the absorbed heat could be determined. Five thickness values, varying from 3.2 to 9.56 mm, were used. These thickness range limits were chosen to allow full penetration of the thinnest plate, as much to prevent equipment damages due to the sample dropping weight (thickest plate).

Figure 6.8 presents the trends related to the thickness influence on the calorimetric measurements. The fitted curve and the extrapolated heat input values, without and with linearization, are shown (statistics are presented in Table 6.3). It can be seen that thicker plates lead to higher absorbed heat. This behavior can be explained based on the fact that both thin and thick plates lose heat through the top surface (radiation and natural convection) and a back surface (natural convection). However, heat loss by radiation from the back surface also happens with thinner plates (full-penetrated welds case). Higher surface temperatures are reached with the thinnest plate, as observed in Figure 6.9. This temperature is high enough to justify radiation loss and higher natural convection losses.

If an extrapolation to a null weld length is applied to all thickness fitted curves, as seen in Figure 6.8, the absorbed heat per unit of bead length reaches the same value. The explanation to these results is the principle of the gross heat input concept, which is the amount of heat determined before

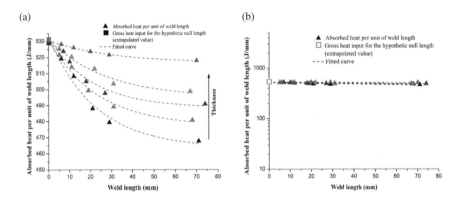

**FIGURE 6.8**
Heat absorbed by the test plate from welds at different lengths for various test plate thicknesses and respective fitted curves (mean welding energy of 621 J/mm, plate dimensions of
$200 \times 100$ and fall distance of 420 mm): (a) the fitted curves showing a natural exponential "e"
behavior and (b) the same curves after a log-linearization (Liskevych and Scotti, 2015). (With
kind permission from Elsevier.)

**TABLE 6.3**

Extrapolated Gross Heat Input Values and Statistics, Related to Figure 6.5

| Test Plate Thickness (mm) | Gross Heat Input (J/mm) | Fit Curve Equation | DF | R² | F | StE |
|---|---|---|---|---|---|---|
| 3.2 | 532.6 | $y = 464.6 + 67.2e^{-x/20.7}$ | 2 | 0.971 | $3.6 \times 10^{-7}$ | 0.7 |
| | 529.2 | $y = 529.2 - 0.2x$ | 3 | 0.886 | $2.0 \times 10^{-6}$ | 1.2 |
| 4.75 | 531.5 | $y = 477.2 + 53.7e^{-x/22.8}$ | 2 | 0.982 | $8.9 \times 10^{-7}$ | 1.1 |
| | 530.7 | $y = 530.7 - 0.5x$ | 3 | 0.884 | $1.7 \times 10^{-6}$ | 2.0 |
| 6.3 | 532.5 | $y = 488.4 + 43.2e^{-x/21.9}$ | 2 | 0.950 | $2.1 \times 10^{-7}$ | 0.9 |
| | 531.4 | $y = 531.4 - 0.5x$ | 3 | 0.887 | $1.3 \times 10^{-7}$ | 1.5 |
| 7.95 | 531.9 | $y = 495.4 + 35.1e^{-x/29.6}$ | 2 | 0.985 | $1.5 \times 10^{-8}$ | 0.7 |
| | 531.0 | $y = 531.0 - 0.7x$ | 3 | 0.883 | $7.7 \times 10^{-6}$ | 1.0 |
| 9.56 | 531.1 | $y = 516.8 + 13.7e^{-x/28.3}$ | 2 | 0.997 | $8.3 \times 10^{-8}$ | 1.2 |
| | 530.2 | $y = 530.2 - 0.8x$ | 3 | 0.884 | $4.1 \times 10^{-6}$ | 1.5 |

*Note:* DF, degree of freedom of the regression; $R^2$, data fitting correlation; $F < 0.05$ means that
the estimator is statistically significant at 95% of confidence level; StE, predicted standard
error of the estimator.

dissipating into the plate and loss through the surfaces. It means that the
intrinsic error due to the test plate thickness vanishes if the gross heat input
is determined by extrapolating absorbed heat values to a bead length tending
to 0. However, it is important to mention that this latter statement is true only
if the effect of another intrinsic error is constant, as, for instance, the present
case of the fall distance. In the case that the test plate width, for instance, is

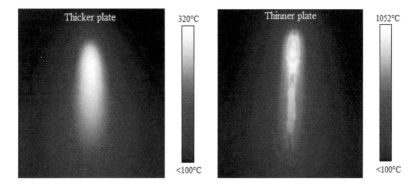

**FIGURE 6.9**
Infrared thermo-graphic maps carried out on the backside of the plates during partial penetration welding (left—9.56 mm thick) and full penetration welding (right—3.2 mm thick): notice that the temperature scales on the right side are different from each other (Liskevych and Scotti, 2015). (With kind permission from Elsevier.)

not wide enough (to be discussed ahead), its effect on the measured absorbed heat values would be different for each thickness. Consequently, there would not be convergence in the extrapolation of all thickness fitted curves to the same absorbed heat per unit of bead length.

### 6.4.3.2 Error due to Test Plate Width

If the test plate were wide enough, one could expect that the edges would turn hot before dropping the plate into the flask (no intrinsic error, since the plate, would have thermally infinite width). Therefore, to determine the limiting width, seven experimental runs were carried out with the test plate width varying from 40 to 110 mm. As suggested before, the flask bottleneck size limited the maximum width dimension. In turn, the minimum width was arbitrarily chosen to prevent burn-through of the weld. Two plate thicknesses (3.2 and 9.56 mm) were used. Weld bead lengths were varied from 5 to 70 mm (in total six steps) for each thickness and width of the test plates, always keeping the same welding parameters.

The curve fitting and the extrapolation to find the gross heat input values are presented in Figures 6.10 and 6.11, respectively (statistical data are provided in Tables 6.4 and 6.5). The same trends are observed for a given test plate thickness, i.e., the wider the test plate, the higher the value of the heat absorbed per unit of weld length. The detected tendencies are reasonably explained for narrower plates by the fact that the lateral plate edges reach higher temperatures before having the plate released into the cryogenic flask due to the heat conduction inside the test plate (Figure 6.12). It should be noted that hot edges lead to extensive higher heat loss by natural convection. Therefore, absorbed heat for the narrower plates is always lower.

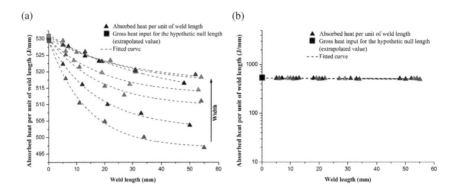

**FIGURE 6.10**
Heat absorbed by the 9.56-mm-thick test plates with different widths as a function of bead length and respective fitted curves (mean welding energy of 621 J/mm, plate length 200 mm and fall distance of 420 mm): (a) the fitted curves showing a natural exponential "e" behavior and (b) the same curves after a log-linearization (Liskevych and Scotti, 2015). (With kind permission from Elsevier.)

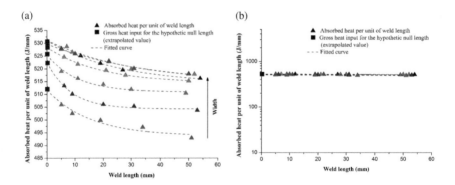

**FIGURE 6.11**
Heat absorbed by the 3.2-mm-thick test plates with different widths as a function of bead length and respective fitted curves (mean welding energy of 621 J/mm, plate length 200 mm and fall distance of 420 mm): (a) the fitted curves showing a natural exponential "e" behavior and (b) the same curves after a log-linearization (Liskevych and Scotti, 2015). (With kind permission from Elsevier.)

Nevertheless, there is the tendency for the extrapolated gross heat input values to converge only for wider plates, i.e., wider than 90 mm for the given welding conditions. The extrapolated values of the gross heat input are very similar (530 ± 1 J/mm) as long as the test plate is wide enough (≥90 mm). Naturally, this pointed width limit is dependent on the material properties and welding energy (for infinitely wide plates, there is no intrinsic error due to plate width). At this point, it could be debatable that absorbed heat from any plate width should extrapolate to a single value at 0 weld length, as happened with plate thicknesses. However, this different behavior (as seen in Figure 6.11) does not

**TABLE 6.4**

Extrapolated Gross Heat Input Values and Statistics, Related to Figure 6.10 (9.56-mm-Thick Test Plate)

| Test Plate Width (mm) | Gross Heat Input (J/mm) | Fit Curve Equation | ANOVA | | | |
|---|---|---|---|---|---|---|
| | | | DF | $R^2$ | F | StE |
| 110 | 530.7 | $y = 508.6 + 21.1e^{-x/46.5}$ | 2 | 0.978 | $9.5 \times 10^{-7}$ | 1.3 |
| | 529.5 | $y = 529.5 - 0.3x$ | 3 | 0.991 | $2.1 \times 10^{-6}$ | 0.9 |
| 100 | 530.9 | $y = 517.2 + 15.7e^{-x/20.0}$ | 2 | 0.948 | $2.5 \times 10^{-6}$ | 1.5 |
| | 530.5 | $y = 530.5 - 0.2x$ | 3 | 0.887 | $8.2 \times 10^{-6}$ | 0.9 |
| 90 | 531.4 | $y = 517.6 + 13.8e^{-x/20.8}$ | 2 | 0.949 | $1.7 \times 10^{-6}$ | 0.9 |
| | 530.7 | $y = 530.7 - 0.2x$ | 3 | 0.989 | $8.4 \times 10^{-6}$ | 1.4 |
| 80 | 529.4 | $y = 512.9 + 16.6e^{-x/20.3}$ | 2 | 0.984 | $2.4 \times 10^{-9}$ | 1.3 |
| | 529.5 | $y = 529.5 - 0.3x$ | 3 | 0.894 | $1.3 \times 10^{-6}$ | 0.8 |
| 70 | 529.7 | $y = 509.3 + 20.5e^{-x/18.7}$ | 2 | 0.970 | $8.2 \times 10^{-9}$ | 1.4 |
| | 528.9 | $y = 528.9 - 0.3x$ | 3 | 0.887 | $7.7 \times 10^{-6}$ | 1.2 |
| 60 | 528.6 | $y = 502.3 + 26.3e^{-x/18.3}$ | 2 | 0.991 | $1.0 \times 10^{-6}$ | 1.1 |
| | 528.5 | $y = 528.5 - 0.5x$ | 3 | 0.994 | $6.0 \times 10^{-6}$ | 0.7 |
| 40 | 528.2 | $y = 496.6 + 31.7e^{-x/14.4}$ | 2 | 0.989 | $1.6 \times 10^{-6}$ | 1.3 |
| | 528.0 | $y = 528.0 - 0.5x$ | 3 | 0.983 | $3.5 \times 10^{-6}$ | 1.6 |

*Note:* DF, degree of freedom of the regression; $R^2$, data fitting correlation; $F < 0.05$ means that the estimator is statistically significant at 95% of confidence level; StE, predicted standard error of the estimator.

**TABLE 6.5**

Extrapolated Gross Heat Input Values and Statistics, Related to Figure 6.11 (3.2-mm Thick Test Plate)

| Test Plate Width (mm) | Gross Heat Input (J/mm) | Fit Curve Equation | ANOVA | | | |
|---|---|---|---|---|---|---|
| | | | DF | $R^2$ | F | StE |
| 110 | 531.0 | $y = 513.7 + 17.3e^{-x/27.1}$ | 2 | 0.984 | $7.2 \times 10^{-7}$ | 1.1 |
| | 531.2 | $y = 531.2 - 0.8x$ | 3 | 0.889 | $1.8 \times 10^{-6}$ | 1.1 |
| 100 | 530.8 | $y = 512.3 + 16.5e^{-x/21.6}$ | 2 | 0.960 | $1.7 \times 10^{-6}$ | 0.8 |
| | 530.7 | $y = 530.7 - 0.2x$ | 3 | 0.989 | $5.6 \times 10^{-6}$ | 1.0 |
| 90 | 529.9 | $y = 514.5 + 15.5e^{-x/33.2}$ | 2 | 0.993 | $3.7 \times 10^{-7}$ | 1.4 |
| | 530.2 | $y = 530.2 - 0.3x$ | 3 | 0.991 | $2.8 \times 10^{-6}$ | 0.8 |
| 80 | 527.9 | $y = 513.9 + 13.9e^{-x/20.9}$ | 2 | 0.998 | $3.5 \times 10^{-7}$ | 1.2 |
| | 528.1 | $y = 528.1 - 0.3x$ | 3 | 0.992 | $6.0 \times 10^{-6}$ | 0.6 |
| 70 | 524.6 | $y = 510.6 + 14.0e^{-x/11.8}$ | 2 | 0.984 | $1.4 \times 10^{-8}$ | 1.0 |
| | 525.1 | $y = 525.1 - 0.3x$ | 3 | 0.884 | $1.1 \times 10^{-6}$ | 0.9 |
| 60 | 521.6 | $y = 504.1 + 17.5e^{-x/8.9}$ | 2 | 0.972 | $9.8 \times 10^{-7}$ | 1.5 |
| | 520.9 | $y = 520.9 - 0.3x$ | 3 | 0.887 | $4.5 \times 10^{-6}$ | 2.1 |
| 40 | 507.3 | $y = 487.1 + 20.3e^{-x/43.2}$ | 2 | 0.948 | $3.1 \times 10^{-6}$ | 1.3 |
| | 508.2 | $y = 508.2 - 0.4x$ | 3 | 0.991 | $5.0 \times 10^{-6}$ | 1.7 |

*Note:* DF, degree of freedom of the regression; $R^2$, data fitting correlation; $F < 0.05$ means that the estimator is statistically significant at 95% of confidence level; StE, predicted standard error of the estimator.

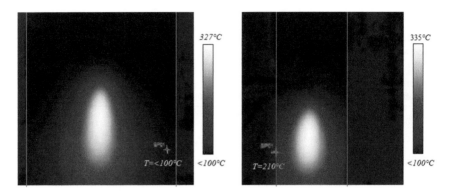

**FIGURE 6.12**
Infrared thermographic maps carried out on the backside of the 9.56-mm-thick test plates during welding with wider (left—100 mm width) and narrower (right—60 mm width) plates; the plate edge is defined by the overlapped vertical lines (Liskevych and Scotti, 2015). (With kind permission from Elsevier.)

mean that the gross heat input determination was misled. As claimed in the start point of this section, the actual gross heat input can be reached only if all intrinsic errors are fully eliminated. When an infinite-like dimension plate is used, as in Figure 6.10, there is still the elapsed time and trajectory intrinsic error. Consequently, an analog behavior seen in Figure 6.11 would also happen in the thickness effect verification if each measurement of the absorbed heat were carried out with a not wide enough test plate.

On the other hand, if the test width to be applied is below the mentioned width limit, for instance, in the case in which the maximum width allowed by the calorimeter is smaller than the width limit, an extrapolation process analog to the treatment given to the fall distance must be applied to find the gross heat input. Unfortunately, a limited number of experiments would not be sufficient to have a real tendency curve fitting in this case. As an alternative, a correspondent generic curve should be obtained to predict heat losses from the plate edges. In this case study, the Rosenthal model could be applied. Rosenthal's model allows obtaining the transversal distance from the heat source for any temperature applied (temperature on the plate edge), as illustrated in Figure 6.13a. Furthermore, some experimental results should be applied over this generic curve to uncover the parametric indices and to determine the fitting line for the given welding conditions, as seen in Figure 6.13b. Thus, both the infinite width and the correspondent gross heat input can be determined. Accordingly, with this method, an infinite width was estimated as being of 92 mm, which is in full agreement with the experimental results.

### 6.4.3.3 Error due to Test Plate Length

Heat can be lost from the test plate end surfaces before the calorimeter measurement according to the distance from the edges that the weld bead

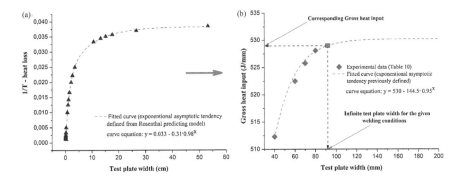

**FIGURE 6.13**

(a) Fitted curve equation defined from the Rosenthal predicting model, where $T$ means the reached temperature at the plate edge (the higher $T$, the higher the heat loss through the plate edges) and (b) extrapolation of gross heat input values for different test plate widths to a width tending to infinite (supposing that the calorimeter limit width is 92 mm) (Liskevych and Scotti, 2015). (With kind permission from Elsevier.)

starts and finishes. In this context, the bead starting to position deserver more attention, because of the typical isotherm shapes from a moving arc (elongates backward). In this work, this distance was left as 50 mm. To verify if this distance would be long enough so as not to induce an intrinsic error, welds were carried out and calorimetric measurements applied under variation of the bead starting position at 50, 75, and 100 mm (maintaining the same welding parameters, test plate dimension, and bead length). According to the results for these given test conditions, no effect of the bead starting point on the absorbed heats was detected (no correction needed to the gross heat input). However, in the case of the influence of this parameter, the determination of the gross heat input at each starting distance should be implemented. This can be done with a series of welding with different bead lengths, beginning at progressively crescent distances from the plate edges (extrapolation of the fitted curves of the absorbed heat to a bead length tending to 0). The gross heat input values could be reached as a function of the bead starting distances to a bead starting distance tending to infinite.

As proposed above, and detailed in the original bibliography source (Liskevych and Scotti, 2015), the actual gross heat input value can be determined experimentally, even though at time and cost expenses. This methodology, when applied to liquid nitrogen calorimetry, can be summarized as follows:

a. To determine the absorbed heat from a welding condition (process/ parameters) in an automated liquid nitrogen calorimeter for different and progressively increased bead lengths (e.g., 5, 10, 20, 35, 55, and 85 mm long) using a given fall distance and test plates as thick, long, and wide as possible, according to the calorimeter dimensions and load capacity;

b. To plot the absorbed heats as a function of bead length and extrapolate its value to a null bead length (reaching the gross heat input, still potentially contaminated by other intrinsic errors);

c. To determine the absorbed heat from at least three other distance falls (increasing from the minimum distance) and different bead lengths for determining the error due to elapse time-distance fall by plotting the absorbed heat as a function of bead length and extrapolating (preferably with linearization) the value to a null bead length;

d. To determine the gross heat input from a linearization of the absorbed heat data from different distance falls and extrapolate its value to a null distance fall (reaching the gross heat input, still potentially contaminated by remaining intrinsic errors);

e. To determine the absorbed heat from at least four decreasing plate widths (from the maximum width allowed by the equipment) and a given bead length for determining the minimum (limit) test plate width for the given process/parameter and plate properties condition;

f. If the test plate width used in "a" can be considered infinite from the thermal equilibrium point of view, there is no intrinsic error to deal with. Otherwise, in an analogous way to items "c" and "d", to determine the intrinsic error due to test plate width;

g. To determine the absorbed heat from at least three increasing bead starting distances from the plate end and a given bead length for determining the minimum (limit) test plate bead starting distance for the given process/parameter and plate property condition;

h. If the bead starting position used in "a" can be considered infinite, there is no intrinsic error to deal with. Otherwise, the error must be determined, in a similar way to items "c" and "d", to determine the intrinsic error due to bead starting distance;

i. The actual gross heat input must be the one calculated in "c" added to the determined errors from steps "f" and "h" if needed.

It is import to note that the proposed methodology was verified only for an automated cryogenic calorimeter. However, considering the theoretical background applied (intrinsic error determination and its expurgation), one cannot foresee any reason to not be applied to any other calorimetric method. One must be aware that the methodology for the actual gross heat input determination as proposed above is rather complex and laborious (due to the need for sophisticated equipment and application of extrapolation techniques). Therefore, one could argue that this approach would not be the best option for ordinary engineering applications, yet it would be of interest for developing numeric predicting models and for welding research that always demands precise heat input values.

## 6.5 Which Is the More Adequate Representation of Current When Heat Input into the Plate Is Concerned, Though Mean or RMS Values?

It is clear for most welding engineers that heat input governs bead formation. However, in addition to the difficulty of determining heat input, as revealed above, the importance of the methods to calculate welding energy in arc welding is hardly observed in the scientific studies. There are different approaches to determine arc power, namely, arithmetical average power (Equation 6.1), effective average power (Equation 6.2), and average instantaneous power (Equation 6.3).

$$P_{\text{arithmetical}} = U_m \cdot I_m \tag{6.1}$$

$$P_{\text{effective}} = U_{\text{ef}} \cdot I_{\text{ef}} \tag{6.2}$$

$$P_{\text{inst}} = \frac{1}{n} \sum_{k=1}^{n} U(k) \cdot I(k) \tag{6.3}$$

where $U_m$ and $I_m$ are the average of voltage and current measured values, $U_{\text{ef}}$ and $I_{\text{ef}}$ are the voltage and current RMS (root-mean-square) values, $U(k)$ and $I(k)$ are the instantaneous voltage and current measured values and where $n$ is the number of samples.

Unfortunately, there is no consensus among authors regarding the methods that should be used for calculating arc energy. However, Bosworth (1991) has found that the differences of applying the different methods on the final value can reach 30%. Joseph et al. (2003), using a calorimetry, stated that the only measure of welding energy which is reasonably well correlated to current variations is based on the instantaneous power. Nascimento et al. (2007) analyzed all the methods mentioned above and the respective consequences on the heat input and thermal efficiency calculations. They claimed that the arithmetic means power method can be applied in a few cases, in which there is no oscillation in current and voltage (like in spray transfer gas metal arc welding), but it is safer to use the average instantaneous power method (more laborious, yet generic). They supported their claim based on the expectancy theorem, which says that the product of the means of two discrete variables is the same as the mean of the product of these variables only if the variables are independent of each other. Melfi's (2010) paper shows that the method to calculate heat input, using average instantaneous energy, was added in the 2010 edition of the ASME Boiler and Pressure Vessel Code: Section IX (item QW409.1) for waveform-controlled welding. The ISO/TR 18491:2015 standard is also using this method in the guidelines for measurement of welding energies. The concept of Equation 6.3 is the integral of power (different from Equations 6.1 and 6.2), and thus applicable to any current (DC and AC) and welding conditions (from steady constant current to unsteady pulsed current). Thus, average instantaneous power seems to be gaining acceptance in the welding community.

Nevertheless, the role that currently plays in the arc energy, consequently on heat input, is yet more obscure. Silva and Scotti (2017) described that Lesnewich (1958), several years ago, demonstrated experimentally that the wire melting rate (MR) in free-flight transfer GMAW can be expressed by a generic and empirical equation, as reproduced in Equation 6.4. This equation has been used years after years by scholars and welding engineers for estimating welding parameters for particular conditions:

$$MR = \alpha I + \beta L I^2 \tag{6.4}$$

where MR is the wire melting rate, $I$ is the current, $L$ is the electrode length (electrified electrode extension), and $\alpha$ and $\beta$ are the constants (which is the function of polarity, shielding gas, and electrode material and diameter).

Lesnewich's equation passed through some improvements in recent years, considering technical limitations at the time that this equation had been elaborated. If MR, for instance, is integrated, a new representation of Equation 6.4 is obtained:

$$\overline{MR} = \frac{1}{T} \int_0^T MR.dt, \tag{6.5}$$

where $T$ represents the period. Replacing MR in Equation 6.5 by Equation 6.4 gives

$$\frac{1}{T} \int_0^T (\alpha i + \beta L i^2).dt \quad \Rightarrow \quad \alpha \frac{1}{T} \int_0^T i\, dt \quad + \quad \beta L \frac{1}{T} \int_0^T i^2\, dt \tag{6.6}$$

where $i$ represents the instantaneous value of current. Mean current and effective current (RMS) are defined as follows:

$$\frac{1}{T} \int_0^T i\, dt = I_m, \tag{6.7}$$

$$\frac{1}{T} \int_0^T i^2\, dt = I_{rms}^2, \tag{6.8}$$

where $I_m$ stands for the mean value of current and $I_{rms}$ stands for the RMS value of current. Substituting Equations 6.7 and 6.8 into Equation 6.6 yields Lesnewich's equation as follows (i.e., both mean and RMS values of currents are used to quantitatively define MR):

$$MR = \alpha I_m + \beta l I_{rms}^2 \tag{6.9}$$

Nevertheless, Equation 6.9 cannot be applied to all metal transfer modes. In short-circuiting GMAW, Equation 6.9 can estimate the electrode MR only during the period of arcing. During the short-circuit period, MR equation should consider only the Joule effect (no cathodic coupling anymore). Therefore, an upgraded version of Equation 6.4 for short-circuit transfer can be written as follows:

$$MR = \alpha I_m + \beta l I_{rms}^2 \Big|_0^{tarco} + \beta l I_{rms}^2 \Big|_{tarc}^{tcc}. \qquad (6.10)$$

Bead formation also takes current to be formulated by models, even though the phenomena involved are much more complex than the ones of MRs. Gunaraj and Murugan (2000) on submerged arc welding, Nagesh and Datta (2002) on shielded metal arc welding, Alam and Khan (2011) on submerged arc welding, Singh et al. (2012) on shielded metal arc welding, Ghosh and Hloch (2013) on submerged arc welding, and Sudhakaran et al. (2013) on gas tungsten arc welding are some of the publications in which geometry modeling was the target. Experimental statistics were applied in all these pieces of work. It is important to mention that mean values were the only representation of current in these researches, i.e., bead geometry is presented as a function of $I_m$ (together with other parameters, such as contact tip-to-work distance and travel speed). However, Mishra and DebRoy (2005) mention that welding systems are highly complex and involve the nonlinear interaction of several welding variables to justify a restricted use of advanced numerical heat transfer and fluid flow codes for fusion welding. Consequently, differently, from MR, bead formation is governed by a complex and interconnected relationship between thermal effects (heat generated on the plate surface) and mechanical effects (forces action on the pool surface), under different mechanisms. Current is certainly one of the welding variables that governs both thermal and mechanical effects, as schematically shown in Figure 6.14.

However, as seen in the current literature, almost nothing has been published about bead geometry modeling, in which current is represented by RMS values. Omar and Lundin (1979) was the first paper to the knowledge of

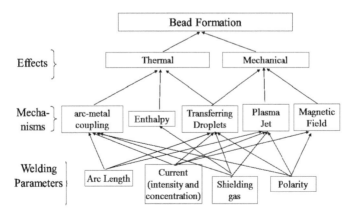

**FIGURE 6.14**
Inter-correlation of variables and mechanisms that govern weld bead formation (Silva and Scotti, 2017). (With kind permission from Elsevier.)

this author that dealt directly with the current representation either by mean or by RMS values. They concluded that, in general, the average current is the controlling welding parameter on melting efficiency for both the PAW and GTAW processes. The second paper with this same purpose was recently published by Cunha et al. (2016), who employed both different values of pulse amplitude to reach a same mean current with different RMS values and variations of pulse amplitude values to obtain different mean current values without changing the value of the RMS current. Cunha et al.'s results evidence that the weld penetration behavior is closely related with mean welding current, while the weld width is also related to the RMS value of the welding current.

Mishra and DebRoy (2005) demonstrated that specified weld geometries can be reached from a computational procedure using a genetic algorithm based on the conservation of mass, momentum, and energy equations. Only mean values of current were also used by them to define the welding parameters as input. However, in numerical modeling, bead geometry can also be modeled according to current variation with time (instantaneous values), and not according to average values (such as $I_m$ or $I_{rms}$). Kim and Na (1998), for instance, predicted that the width and depth of the weld pool increase steeply during peak current period when pulsed TIG was simulated. The welding conditions applied by the authors in their simulations accounted for pulse parameters arranged in such a way that $I_{rms}$ was kept constant and $I_m$ was reduced. They claimed that the mean current represents the electrical effects obtained with a constant current of the same value, while the RMS current represents the thermal effects, recognizing that both current representations have roles to play in the bead formation.

Based on the above-presented information and on MR predictors (Equations 6.9 and 6.10), one could suggest that not only mean values would represent current in the weld bead formation, but also, and together, IRM values. It is important to recall that, by representing current either by mean or by RMS values, it does not mean that a similar mean and RMS value would produce the same amount of heat. The definition of RMS related to heat is applicable only to a circuit whose impedance consists of a pure ohmic resistance (known postulated: for a cyclically alternating electric current, RMS is equal to the value of the direct current that would produce the same power dissipation in a resistive load). In a more global manner, RMS refers to the most common mathematical method of defining nonsteady signal waveforms. Therefore, even though knowing that an arc welding circuit (electrode, arc, and plate) would be represented, rather than by an pure resistive circuit, by an *ER* circuit (Jorge et al., 2017), where *R* is the summation of all true time-dependent resistive segments of the circuit and *E* would represent the potential energy to overcome the needs to maintain the arc column and the couplings arc-electrode/droplet and arc-weld pool, RMS values can be determined for DC electrical signals.

Silva and Scotti (2017), aiming to support who develops weld bead modeling, evaluated the weight of mean or RMS values on the current representation of arc welding bead geometry formation. It is important to point out that Silva and Scotti's purpose was not to justify bead formation under welding parameter variations. Rather, the intention was to compare geometries of beads carried out with a given mean current to geometries of beads carried out with this same mean value of current, yet with different RMS values. To do so, the other parameters (travel speed, arc length, electrode diameter, and included angle) were kept constant in this study. The reduction of variables was pursued reducing noises in the result analyses.

The methodology applied by Silva and Scotti (2017) was based on experiments with GTAW, which can be used either with constant and pulsed DCEN current. With a constant current, both mean and RMS values are the same (when current output is straight, possible with electronic power sources). However, the RMS values can be different even keeping the same mean values. This is possible if GTAW is used with pulsed current and deliberated levels and duration of the pulse and base are set, as illustrated in Figure 6.15. It is important to mention that although the nominal values of $I_{rms}$ become progressively greater than $I_m$ values as $\Delta I$ is increased, it does not mean that the first representation of current is providing more heat to the plate, since mean and RMS are only different ways of denoting the same current as a function of time.

Following this concept, Silva and Scotti's (2017) developed an experimental design with six combinations of pulse-based current that made the mean current values equal to 200 A, i.e., with $\Delta I$ of 10, 20, 50, 100, 200, and 300 A. In this parametrization, pulse and base durations were strategically chosen (tp = tb = 0.25 s), to provide 2 Hz of pulsing frequency. These settings led the RMS current values from 200 to 249 A. In the same experimental design,

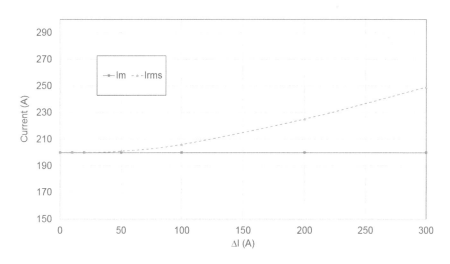

**FIGURE 6.15**

Example of the relationship between RMS and mean values of current for increasing $\Delta I$ (pulse–base current) at a same mean current (Silva and Scotti, 2017). (With kind permission from Elsevier.)

another set of four welds with constant currents was carried out with the mean values the same as the RMS values, i.e., $I_m = I_{rms} = 200, 206, 224$, and 249 A. Figure 6.16 presents the effect of the current expressed as both mean and RMS values on bead penetration, bead width, and fused area. Two facts are distinguished in Figure 6.16. There exist abnormal results (outliers) when the weld with pulsed-base current with $\Delta I = 100$ was used. To confirm it, the test was replicated, suggesting an uncovered effect for this condition (likely a phenomenon related to a particular vibration mode of the pool at this combination of pulse amplitude and frequency). The second fact is that there are no significant differences between mean and RMS values of current when welding with small $\Delta I$ values (10, 20, and 50) in pulsed mode. These conditions were planned to confirm the trends toward higher $\Delta I$ values (confirming the behavior for $\Delta I \leq 100$ A), but the outcomes also confirmed the reliability of the experiments, since they were statically the same as $\Delta I = 0$ for $I = 200$ A.

On the left side of the vertical line on the charts of Figure 6.16 are the results from the welds with constant current (i.e., same values of mean and RMS currents). As can be seen from the figure, deeper penetration, larger fused area, and wider beads happen as current is increased. These trends were expected since arc energy also proportionally increased (potentially higher heat input). To confirm these trends, one-way ANOVA was applied on the data ($I_m = I_{rms}$ as the only factor and penetration, bead width, and fused area as separate responses). The resultant $p$-values of 0.000, 0.001, and 0.000, respectively, point out that the rise of $I_m = I_{rms}$ presents high significance relation to the increase of the responses at a level of 0.005 and a total degree of freedom of 11. Taking now the right side of the vertical line on the charts of Figure 6.16, where mean current is kept constant and RMS values are increased, the tendencies are of penetration, fused area, and bead width to remarkably increase as RMS values are augmented (especially when the outliers at $\Delta I = 100$ are removed). The sensitivity scale (gradient) is higher on the right side for penetration and lower for fused area and bead width. By applying again one-way ANOVA on the data ($\Delta I$ as the only factor and penetration, bead width, and fused area as separate responses), the resultant $p$-values of 0.000, 0.015, and 0.007, respectively, indicate that the factor $\Delta I$ is significant on each response at a level of 0.005 and a total degree of freedom of 14.

It is worth mentioning that increasing RMS values at the same $I_m$ did not affect penetration, according to results of Omar and Lundin (1979). However, these authors worked with a much lower mean current ($I_m = 50$ A) than that used in the work of Silva and Scotti (2017), varying $I_p$ and $I_b$ to reach $I_{rms}$ from 54 to 64 A (a difference for mean values of maximum 14 A). As seen in Figure 6.16a, the effects of RMS values on penetration in Silva and Scotti's work were significant only when the differences to mean values were over 24 A. In addition, Omar and Lundin varied $I_p$ and $I_b$ to reach $I_{rms}$ from 54 to 64 A (a difference for mean values of maximum 14 A).

Apparently, bead formation, based on these results, would be better represented by RMS values of current, since only RMS values increased at both

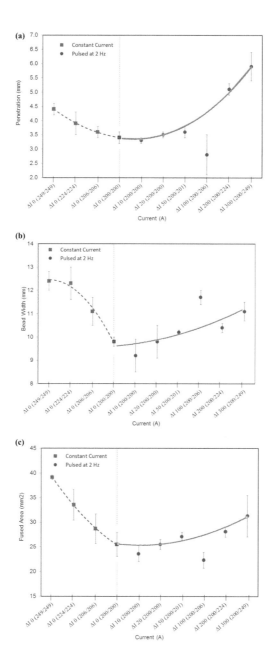

**FIGURE 6.16**
Influence of mean and RMS currents on the bead penetration (a), bead width (b), and fused area (c) (averages and standard deviations of three cross sections from a same coupon test); codification: $\Delta xxx$ (yyy/zzz), where "xxx" is the difference between pulse and base currents, "yyy" the mean current and "zzz" the RMS current): the vertical line split the parametric conditions between constant current (left) and pulsed current (right) (Silva and Scotti, 2017). (With kind permission from Elsevier.)

sides of the figures (mean values were constant at the right side). However, RMS values alone cannot represent the bead formation because when RMS values are increased together with mean values (left side), a less-sensitive increase of penetration is noted. For fused area and bead width, the trends showed a steeper gradient on the left side. As mentioned earlier, the complex parameter relation in bead formation justifies the different weight of the current representation for each of the geometric parameters.

In a complementary way, a comparison between parametric conditions with the same RMS values yet with different mean values was also possible with the planned experimental design. If one compares steady data from the right and left sides ($\Delta I$ 300 (200/249)) $\times$ ($\Delta I$ 0 (249/249)) and ($\Delta I$ 200 (200/224)) $\times$ ($\Delta I$ 0 (224/224)), it can be inferred that the increase of the mean values, yet keeping the same RMS values, also influenced bead penetration, bead width, and fused area. In Figure 6.16b,c, fused area and bead width grew up as mean current increases. Conversely, looking at Figure 6.16a, the penetration gets shallower if higher values of mean current are reached (left side) at the same RMS values, yet with no decrease or even a slight increase for low mean values.

According to Omar and Lundin (1979), the outcomes on penetration are in agreement with the ones of the current work if only data with low differences between mean and RMS values are compared. Omar and Lundin showed that penetration increased around 0.5 mm when mean values increased from 40 to 47 A for an $I_{rms}$ of 50 A. Kim and Na's (1998) outcomes originated from numerical simulation over pulsed TIG data, in turn, were totally in concordance with the results of Silva and Scotti's (2017) work, i.e., the increase of mean current for the same RMS current led to wider and swallower beads. In their simulation, the increase of $I_m$ (85–92 A) for the same $I_{rms}$ (100 A) was obtained by pulsing the TIG arc with the same frequency (proximately 2 Hz), yet $\Delta I$ was decreased from 112 to 84 A and pulse time ratio (pulse current duration/base current duration) from 0.5 to 2. These results reinforce the suggestion that both mean and RMS values represent the phenomena. On the other hand, Cunha et al. (2016) found a close relationship between mean welding current and the weld penetration behavior, although the weld width is related with the RMS value of the current signals. The latter authors worked with higher mean and RMS currents (150 A) than Kim and Na (1998) and Omar and Lundin (1979), yet lower than in Silva and Scotti's (2017) study. Another difference between the experimental conditions above presented and Cunha et al.'s is the fact the pulsed frequency here was of only 2 Hz, in contrast to 50 Hz used by Cunha et al. The similarities and mismatches of the three results do not invalidate the hypothesis that both mean and RMS values would also represent current in the weld bead formation. On the contrary, they show only that the weight of each representation differs according to the welding parameters.

Silva and Scotti (2017) still proposed new experiments (with same settings as in the previous sections), targeting to expand the basis for analysis on the representability of mean and RMS values on bead formation. They measured the heat transferred to the plate and correlated this quantity to current, represented by both Mean and RMS values. Although Hurtig et al. (2016) and Liskevych et al. (2013) reported limitations of calorimetric methods for heat input measurement, a cryogenic calorimeter, as described by Liskevych and Scotti (2015), was used. Differently from recommended in this latter citation, to save resources all comparisons were carried out following just one test setting condition, i.e., automated liquid nitrogen calorimeter operation, fixed plate trajectory, and bead lengths. Considering to be impossible of avoiding, yet feasible of mitigation, the test plates were large enough and of unaltered dimensions to face the intrinsic errors due to heat losses before measurement. Even after these precautions, a lower absorbed heat than expected may have been reached as a consequence of higher heat losses before measurement when heat transfer was big. In summary, the study of the weight of mean and RMS values of current was made on the trends (and not on absolute values) of the absorbed heat during the cryogenic calorimetric test (which is not necessarily the heat input).

The heat absorbed by the test plates as a function of current (both mean and RMS values) is presented in Figure 6.17. As can be seen in the figure, the trends are clearer than for geometric parameters and data performed a smooth trend line, even considering just one test per parametric condition. In addition, the condition ($\Delta I = 100$) that had presented abnormal behavior in the figures related to penetration, width, and area did not present the same behavior. This happening is fairly explicable since the thermal part of the energy is what governs the heat transferred and absorbed by the plate (bead formation accounts for thermal and mechanical energies) in calorimetric measurements. For mean values of current equal to RMS values (left side of the vertical line) and for increasing RMS values of current keeping the same mean value of current (right side), the increase of current led to an increase of the absorbed heat. The tendency is very similar to those for the fused area in the previous section.

In the attempt for separating the weight of the mean and RMS values on the representation of the absorbed heat, another approach was tried. The pairs of parametric conditions located at the left and right sides of Figure 6.17 (with same RMS values, yet different mean values) were compared, e.g., ($\Delta I$ 300 (200/249)) $\times$ ($\Delta I$ 0 (249/249)). It was established that more heat was absorbed for the higher mean values of current with the similar RMS values, the more the absorbed heat. However, as seen in Figure 6.18, the mean value of current depends on its RMS values. Eventually, it was inferred that for the MR, both values (mean or RMS) are significant for the representation of current (quantified here by the absorbed heat from a cryogenic calorimetric test). The prevalence of the representation of one value over the other may depend on the amplitude of each value. For instance, the slope of the lines in Figure 6.18 suggests that the lower the RMS value, the more predominant is the mean value representability.

Based on the data presented in this work (plain carbon steel GTAW, no material feeding, no joint/groove, pulsing frequency of 2 Hz and current range from 200 to 350 A), it is concluded that models to predict bead geometry can be more precise, at least for GTAW with no material feeding, if both

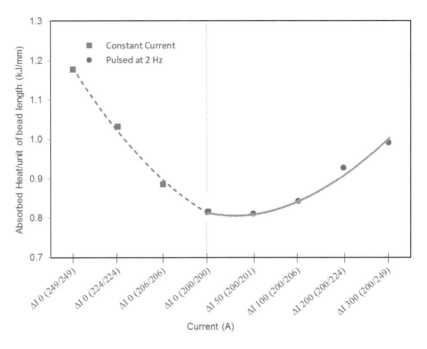

**FIGURE 6.17**
Influence of mean and RMS currents on the absorbed heats; codification: $\Delta xxx$ ($yyy/zzz$), where "$xxx$" is the difference between pulse and base currents, "$yyy$" the mean current and "$zzz$" the RMS current): the vertical line split the parametric conditions between constant current (left) and pulsed current (right) (Silva and Scotti, 2017). (With kind permission from Elsevier.)

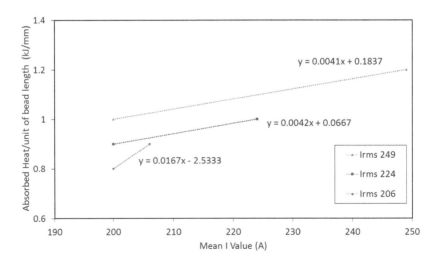

**FIGURE 6.18**
Proportionality ratio of the mean value influence on absorbed heat for different RMS values of current representation (Silva and Scotti, 2017).

mean and RMS values are used to represent current in arc welding bead geometric formation. In other words, bead geometry is a function of both $I_m$ and $I_{rms}$, in addition to other parameters such as contact tip-to-work distance, travel speed, etc.

## 6.6 Summary

Several aspects related to determination of heat input in arc welding and application of its value, either at workshop level or in modeling, were presented and discussed in this chapter. The objective was to bring to discussion the real meaning of heat input, the quality of the determined values and the effectiveness of its application in practice. Initially, a differentiation between heat input and arc energy was established and the importance of this distinction is highlighted. Following, it was demonstrated that a same arc welding energy can deliver different heat inputs, consequently bead geometries, if boundary conditions, such as plate thickness, are changed. A descriptive model of the heat flux inside a plate was presented as a means of understanding the sources of error in existing heat input measurement methods. Calorimetric intrinsic errors were pointed out and methods of error determination were described. The concept of Gross Heat Input was introduced as a way of increasing the usage reliability of heat input values. It is also discussed and shown existing equations for arc power determination, and it was considered the average instantaneous power as the most appropriate means. It was demonstrated that the calorimetric intrinsic errors, added to the inaccuracies of the monitoring devices, can reach figures for arc efficiency calculations as high as ±10%. This quantity would mean that arc efficiency application in modeling is made merely as a calibration factor, with no physical meaning. Gross Heat Input, on the other hand, seems to be a more reliable factor, yet difficult to be determined. Finally, it was concluded that as much as for melting rate and bead formation predicting formulations, current should be represented by both mean and RMS values also for equating out heat input experimentally.

## References

Alam, S., Khan, M. (2011), Prediction of weld bead penetration for steel using submerged arc welding process parameters. *International Journal of Engineering, Science and Technology (IJEST)*, 3(10), pp. 7408–7416.

ANSI/AWS A3.0 (2001), *Standard Welding Terms and Definitions*. Miami, FL: American Welding Society, 152 p.

Arévalo, H.D.H. (2011), Development and assessment of experimental rig for calorimetry via liquid nitrogen and continuous flow (water) in welding processes. *MSc Dissertation*, Federal University of Uberlândia, Uberlândia, Brazil, 145 p. (in Portuguese).

Arévalo, H.D.H., Vilarinho, L.O. (2012), Development and assessment of calorimeters using liquid nitrogen and continuous flow (water) for heat input measurement. *Soldagem e Inspeção*, 17(3), doi: 10.1590/S0104-92242012000300008 (in Portuguese).

Bardin, F., Morgan, S., Williams, S., Mcbride, R., Moore, A.J., Jones, D.C., Hand, P.D. (2005), Process control of laser conduction welding by thermal imaging measurement with a color camera. *Applied Optics*, 44(32), pp. 6841–6848, doi: 10.1364/AO.44.006841.

Bosworth, M.R. (1991), Effective heat input in pulsed gas metal arc welding with solid wire electrodes. *Welding Journal*, 70(5), pp. 111s–117s.

Cângani, A.P.M., Ferraresi, V.A., Guimaraes, G., Carvalho, S.R. (2010), Thermal and experimental analysis of the GTAW process. In: *Inverse Problems, Design and Optimization Symposium, 2010, João Pessoa—PB. Proceedings of the IPDO-2010, August 25–27*, v. 1, Brazil. pp. 1–8.

Cantin, G.M.D., Francis, J.A. (2005), Arc power and efficiency in gas tungsten arc welding of aluminum. *Science and Technology of Welding and Joining*, 10(2), pp. 200–210, doi: 10.1179/174329305X37033.

Carvalho, S.R., Guimarães, G., Silva, S.M.M. (2006), Thermal analysis using inverse technique applied to an aluminum alloy GTAW. In: *International Heat Transfer Conference—IHTC13, Australia. Proceedings of the International Heat Transfer Conference—IHTC13, 2006*, v. 1, Australia. pp. 1–10.

Cunha, T.V., Voigt, A.L., Bohórquez, C.E.N. (2016), Analysis of mean and RMS current welding in the pulsed TIG welding process. *Journal of Materials Processing Technology*, 231, pp. 449–455, doi: 10.1016/j.jmatprotec.2016.01.005.

DuPont, J.N., Marder, A.R. (1995), Thermal efficiency of arc welding processes. *Welding Journal*, 74(12), pp. 406s–416s.

Egerland, S., Colegrove, P. (2011), Using solid state calorimetry for measuring gas metal arc welding efficiency, In W. Sudnik (ed.). *Arc Welding*. London: Intech Open.

Essers, W., Walter, R. (1981), Heat transfer and penetration mechanisms with GMAW and plasma-GMA welding. *Welding Journal*, 60(2), pp. 37s–42s.

Fuerschbach, P.W., Knorovsky, G.A. (1991), A study of melting efficiency in plasma arc and gas tungsten arc welding. *Welding Journal*, 70(11), pp. 287s–297s.

Gery, D., Long, H., Maropoulos, P. (2005), Effects of welding speed, energy input and heat source distribution on temperature variations in butt joint welding. *Journal of Materials Processing Technology*, 167(2–3), pp. 393–401.

Ghosh, A., Hloch, S. (2013), Prediction and optimization of yield parameters for submerged arc welding process. *Technical Gazette*, 20(2), pp. 213–216.

Giedt, W.H., Tallerico, L.N., Fuerschbach, P.W. (1989), GTA welding efficiency: Calorimetric and temperature field measurements. *Welding Journal*, 68(1), pp. 28s–32s.

Gunaraj, V., Murugan, N. (2000), Prediction and optimization of weld bead volume for the submerged arc process—Part 2. *Welding Journal*, 79(11), pp. 331s–337s.

Haelsig, A., Mayr, P. (2013), Energy balance study of gas-shielded arc welding processes. *Welding in the World*, 57(5), pp. 727–734, doi: 10.1007/s40194-013-0073-z.

Haelsig, A., Kusch, M., Mayr, P. (July 2011), A new technology for designation the efficiency of gas shielded arc welding gives new findings. In: *IIW Doc XII-2031-11, Commission XII Meeting, IIW Annual Assembly 2011*, Chennai.

Haelsig, A., Kusch, M., Mayr, P. (2012), New findings on the efficiency of gas shielded arc welding. *Welding in the World*, 56(11–12), pp. 98–104, doi: 10.1007/BF03321400.

Hsu, C., Soltis, P. (2002), Heat input comparison of STT vs. short-circuiting and pulsed GMAW vs. CV processes, In: *6th International Conference Trends in Welding Research, Apr 2002*, Pine Mountain, GA, ASM International, pp. 369–374.

Hurtig, K., Choquet, I., Scotti, A., Svensson, L.-E. (2016), A critical analysis of weld heat input measurement through a water-cooled stationary anode calorimeter. *Journal of Science and Technology of Joining and Welding*, 21(5), pp. 339–350, doi: 10.1080/13621718.2015.1112945.

Jorge, V.L., Gohrs, R., Scotti, A. (2017), Active power measurement in arc welding and its role in heat transfer to the plate. *Welding in the World*, 61(4), pp. 847–856, doi: 10.1007/s40194-017-0470-9.

Joseph, A.P. (2001), Assessing the effects of GMAW-pulse parameters on arc power and weld heat input. *MSc Dissertation*, The Ohio State University, Columbus, OH, 101 p.

Joseph, A., Harwig, D., Farson, D.F., Richardson, R. (2003), Measurement and calculation of arc power and heat transfer efficiency in pulsed gas metal arc welding. *Science and Technology of Welding and Joining*, 8(6), pp. 400–406, doi: 10.1179/136217103225005642.

Kim, W.H., Na, S.J. (1998), Heat and fluid flow in pulsed current GTA weld pool. *International Journal of Heat and Mass Transfer*, 41(21), pp. 3213–3227, doi: 10.1016/S0017-9310(98)00052-0.

Lesnewich, A. (1958), Control of melting rate and metal transfer—Part I. *Welding Journal*, 37(8), pp. 343s–353s.

Lima e Silva, S.M.M., Vilarinho, L.O., Scotti, A., Ong, T., Guimarães, G. (2003), Heat flux determination in the gas-tungsten-arc welding process by using a three dimensional model in inverse heat conduction problem. *High Temperatures–High Pressures*, 35/36(1), pp. 117–126, doi: 10.1068/htjr086.

Liskevych, O., Scotti, A. (2015), Determination of the gross heat input in arc welding. *Journal of Materials Processing Technology*, 225, pp. 139–150, doi: 10.1016/j.jmatprotec.2015.06.005.

Liskevych, O., Quintino, L., Vilarinho, L.O., Scotti, A. (2013), Intrinsic errors on cryogenic calorimetry applied to arc welding. *Welding in the World*, 57(3), pp. 349–357, ISSN 0043-2288, doi: 10.1007/s40194-013-0035-5.

Lu, M.J., Kou, S. (1989), Power inputs in gas metal arc welding of aluminum—Part 1. *Welding Journal*, 68(9), pp. 382s–388s.

Melfi, T. (2010), New code requirements for calculating heat input. *Welding Journal*, 89(6), pp. 61–65.

Mishra, S., DebRoy, T. (2005), A heat-transfer and fluid-flow-based model to obtain a specific weld geometry using various combinations of welding variables. *Journal of Applied Physics*, 98(4), doi: 10.1063/1.2001153.

Nagesh, D.S., Datta, G.L. (2002), Prediction of weld bead geometry and penetration in shielded metal-arc welding using artificial neural networks. *Journal of Materials Processing Technology*, 123(2), pp. 303–312.

Nascimento, A.S., Batista, M.A., Nascimento, V.C., Scotti, A. (2007), Assessment of electrical power calculation methods in arc welding and the consequences on the joint geometric, thermal and metallurgical predictions. *Soldagem e Inspeção*, 12(2), pp. 97–106 (in Portuguese).

Nasiri, M.B., Behzadinejad, M., Lafiti, H., Martikainen, J. (2014), Investigation on the influence of various welding parameters on the arc thermal efficiency of the GTAW process by calorimetric method. *Journal of Mechanical Science and Technology*, 28(8), pp. 3255–3326, doi: 10.1007/s12206-014-0736-8.

Nowacki, J., Wypych, A. (June 2010), Application of thermovision method to welding thermal cycle analysis. *Journal of Achievements of Materials and Manufacturing Engineering*, 40(2), pp. 131–137.

Omar, A.A., Lundin, C.C. (1979), Pulsed plasma—Pulsed GTA arc: A study of the process variables. *Welding Journal*, 58(4), pp. 97s–105s.

Pépe, N., Egerland, S., Colegrove, P., Yapp, D., Leonhartsberger, A., Scotti, A. (2011), Measuring the process efficiency of controlled gas metal arc welding processes. *Science and Technology of Welding and Joining*, 16(5), pp. 412–417, doi: 10.1179/1362171810Y.0000000029.

Quintino, L., Liskevych, O., Vilarinho, L.O., Scotti, A. (2013), Heat input in full penetration welds in gas metal arc welding (GMAW). *International Journal of Advanced Manufacturing Technology*, 68, pp. 2833–2840, doi: 10.1007/s00170-013-4862-8.

Scotti, A., Reis, R.P., Liskevych, O. (2012), A descriptive model of the heat flux in arc welding aiming the effective heat input concept. *Soldagem e Inspeção*, 17(2), 166–172, doi: 10.1590/S0104-92242012000200010 (in Portuguese).

Sikström, F. (2015), Operator bias in the estimation of arc efficiency in gas tungsten arc welding. *Soldagem e Inspeção*, 20(1), pp. 128–133, doi: 10.1590/0104-9224/SI2001.13.

Sikström, F., Hurtig, K., Svensson, L.-E. (2013), *Heat Input and Temperatures in Welding*, 17th JOM—IIW International Conference & Exhibition on the Joining of Materials, May 5–8, 2013, Helsingør.

Silva, D.C.C., Scotti, A. (2017), Using either mean or RMS values to represent current in modeling of arc welding bead geometries. *Journal of Materials Processing Technology*, 240, pp. 382–387, doi: 10.1016/j.jmatprotec.2016.10.008.

Singh, R.P., Gupta, R.C., Sarkar, S.C. (2012), The effect of process parameters on penetration in shielded metal arc welding under magnetic field using artificial neural networks. *International Journal of Application or Innovation in Engineering & Management (IJAIEM)*, 1(4), pp. 12–17.

Soderstrom, E.J., Scott, K.M., Mendez, P.F. (2011), Calorimetric measurement of droplet temperature in GMAW. *Welding Journal*, 90(1), pp. 1s–8s.

Stenbacka, N. (2013), On arc efficiency in gas tungsten arc welding. *Soldagem e Inspeção*, 18(4), pp. 380–390, doi: 10.1590/S0104-92242013000400010.

Sudhakaran, R., Vel Murugan, V., Sivasakthivel, P.S., Balaji, M. (2013), Prediction and optimization of depth of penetration for stainless steel gas tungsten arc welded plates using artificial neural networks and simulated annealing algorithm. *Journal of Neural Computing and Applications*, 22(2013), pp. 637–649, doi: 10.1007/s00521-011-0720-5.

Watkins, A.D. (1989), Heat transfer efficiency in gas metal arc welding. *MSc Thesis*, College of Graduate Studies, University of Idaho, Russia.

Wong, Y.-R., Shih, F.L. (2013), Improved measurement of resistance and calculation of arc power in fusion welding. *Science and Technology of Welding and Joining*, 18(1), pp. 52–56, doi: 10.1179/1362171812Y.0000000074.

Zeemann, A. (2003), Energia de Soldagem, Infosolda. www.infosolda.com.br/artigos/metsol105.pdf (in Portuguese).

# 7

# Friction Stir Welding of High-Strength Steels

**R. Ramesh**
*PSG College of Technology*

**I. Dinaharan and E. T. Akinlabi**
*University of Johannesburg*

## CONTENTS

7.1 Introduction ................................................................................. 139
7.2 Challenges in the Welding of Advanced High-Strength Steel
   Substances ................................................................................... 141
7.3 Comparison to Regular Welding Forms ............................... 141
7.4 Microstructural Zones in FSW ............................................... 143
7.5 Tool Materials ............................................................................ 144
   7.5.1 Refractory Metal Tools ................................................. 144
   7.5.2 Super-Abrasive Tools ................................................... 146
7.6 Tool Wear .................................................................................... 147
7.7 Equipment Requirements for FSW of Steels ........................ 148
7.8 Microhardness Evolution......................................................... 149
7.9 Effect of Process Parameters .................................................. 150
7.10 Microstructural Evolution ....................................................... 151
7.11 Defects in FSW ........................................................................... 154
7.12 Conclusions................................................................................. 155
References.............................................................................................. 155

## 7.1 Introduction

Friction stir welding, *as we know*, has been invented in the year 1991, and its patents are protected by The Welding Institute (TWI) situated in England. It is well known that friction stir welding (FSW) happens to be a solid-state welding procedure. In this method, a turning nonconsumable device will be deciphered along the interface between two materials that are to be merged. The device termed as tool comprises of a projecting pin, and this pin is inserted into the workpiece whereas, a bigger concentric "shoulder" is kept up on the outward part of the joint. The shoulder comprises a sunken surface, which creates an ideal blend of frictional warming and fashioning weight. The schematic setup of FSW is illustrated in Figure 7.1.

These parameters are administered by the tool geometry (i.e., shoulder and pin diameter), mechanical properties of the material to be joined (i.e., flow stress), and material thickness. FSW is primarily used in aluminum alloys. Notwithstanding, there has been a lot of research in the field of higher dissolving temperature materials (HTMs), for example, steel and stainless steel [1–5]. Device materials have been created with sufficient solidness and hardness to withstand the extreme temperature conditions (over 700°C) of FSW. Early FSW applications on ferrous materials, particularly combination steel and stainless steel, were extremely constrained in light of high temperatures that were expected to plastically distort such rigid and sturdy metals and high hot quality required for appropriate device materials to complete tasks of that sort [6].

Therefore, the focal point of FSW has generally been on the low liquefying point metals, for example, Al and Mg combinations because of restricted tool life at high temperatures. Nevertheless, later advances in the tool and process play a vital role in enhancing the performance of this procedure for high liquefying temperature metals like steel and titanium alloys. Recently, various examinations have been carried out concerning different highlights of the FSW of steels and other high-temperature alloys [6]. The FSW process of steel exhibits a variety of advantages crosswise over numerous mechanical segments when compared with ordinary combination-welding methods. Nevertheless, the research about the contact mix welding process in connection to steel remains generally limited [7].

Efforts to apply FSW on Mg alloys and light Al have been started in 1991, and now it is prevalent that it has significant applications in aerospace and transportation fields. Eventually and promptly, numerous researchers initiated studies on FSW of steel and other middling and higher-level melting temperature amalgams such as Cu alloys, stainless steels, Ni-based compounds, and Ti amalgams. From the very beginning, joining and processing such materials were found to be daunting tasks, particularly due to the extreme temperatures and the resultant forces concomitant to the process.

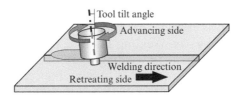

**FIGURE 7.1**
Schematic of FSW setup.

Nonetheless, the impetus for such an endeavor has been motivated by numerous advantages attributed mainly to FSW as follows [8]:

- Owing to its forged microstructure, the mechanical performance of the joints turns out to be good.
- Anticipated lower predisposition to hydrogen embrittlement;
- Lower residual or distortion stresses;
- Higher reproducibility and sturdiness through a lower rate of defect formation;
- Efficient regarding cost and time;
- Intrinsic automation;
- Can weld conservatively unweldable constituents;
- Can weld different types of materials;
- Ensures safety of the welding operator owing to the absence of noxious vapors;
- Innovative prospects for fresh and exceptional applications.

## 7.2 Challenges in the Welding of Advanced High-Strength Steel Substances

1. The very high-carbon and alloying component substances make advanced high-strength steel substances (AHSS) more sensitive to the weld thermal cycle, resulting in a variety of microstructures and properties of the weld.
2. Microstructural features rely largely upon welding conditions and steel chemistry.
3. The welding procedures created for one kind of AHSS may not make a difference to different sorts of materials:
   a. Weld quality
   b. Weld auxiliary execution (static, weakness, effect).

## 7.3 Comparison to Regular Welding Forms

In the hydrocarbon-based fuel sector, the likelihood to deliver exceedingly stable joints including hard to weld materials, such as high-quality steels and joints including unique materials, notwithstanding the possibility to assist and disentangle plant and field joining, has been the primary impetus for improvement in FSW of high dissolving temperature amalgams. Although there are challenges ahead previously, such innovation is grasped by the field workers and utilized enormously inside the petroleum sector [8].

It is known that the atomic materials mandate exceptional criteria of performance and superiority in standards because of their stark surroundings high-energy neutron fission. Oxide scattering reinforced (ODS) ferrite steels have phenomenal opposition to neutron illumination prompted swelling, as well as to high-temperature crawl. In this manner, these compounds are generally thought to guarantee competitor materials for cladding of cutting-edge fast reactors and essential ingredients of fusion reactors. With a specific end goal to apply such amalgams to extensive and confounded structures, joining is an unavoidable and basic preparing stage. They have discovered that ordinary softening welding techniques would bother scattering of finer oxide particles in a compound. As the fantastic crawl opposition and the neutron radiation obstruction of ODS combinations are fundamentally present because of the oxide particles which are usually ultrafine, liquefying welding strategies will not be appropriate and they ought to be replaced by improved welding techniques. FSW, as a solid-state process of fitting together, is thought to be a viable option and favorable approach for welding ODS compounds [9–17].

In automobiles manufacturing, AHSS or advanced high-quality steel sheets have been used to build body structures, and they are additionally anticipated to fulfill the interest of weight reduction. Nevertheless, AHSS sheets have a tendency to possess poor mechanical properties of the weld joint when they are welded with regular forms of welding [18].

Austenitic stainless steel containing a higher level of nitrogen, known as HNS, is one of the predominant supplies that use nitrogen in alloying component as an alternate to the less nifty material of nickel. In this kind of steel, nitrogen is used to improve the development and dependability of austenite and increment the austenite stage run. The high nitrogen level and the attributes of austenitic steels do provide some unusual angles to this material. These incorporate higher quality, high flexibility, and more excellent durability. The high-nitrogen substance can enhance the nearby consumption opposition of stainless steel, particularly setting and cleft erosion obstruction. Moreover, from the perspective of measures to the absence of nickel assets, later on, HNS is being viewed as one of the cutting-edge austenitic stainless steels. Also, high-nitrogen, nickel-less substances in steel maintain a strategic distance from issues with nickel sensitivities.

Naturally, such attributes can lessen the mechanical properties and consumption opposition of welded joints; this again is the reason why HNS is seen as an ineffectively weldable material. In light of such data, use of a strong state joining process for HNS can be performed underneath its softening point.

Reducing the weight of cars and boats is one of the essential approaches to conserve energy and decrease $CO_2$ outflows. Reinforcing constructional materials is one of the conceptual ways to reduce weight. In any case, the more significant part of the primary steel is reinforced by the expansion of precious uncommon metals, for example, Ni and Mo. Furthermore, it is broadly realized that the elasticity of the steel could be increased by increasing the amount of carbon. This implies that the significantly less expensive carbon component can supplant the other expensive metals. Because of the simple event of splits or cracks in the combination-welding zone, the medium- and high-carbon steels, for the most part, are viewed as non-weldable materials [9,11–19].

## 7.4 Microstructural Zones in FSW

There are three viable zones with different material stream and related temperature in FSW. These zones comprise contrasting intensities of material stream as well as related temperature variations. The weld nugget (WN) has higher material flow and temperature, thermo-mechanically affected zone (TMAZ) has lower material flow and moderate temperatures, and heat-affected zone (HAZ) has moderate temperatures but no material flow. The transverse macrostructure of the weld of a steel plate is shown in Figure 7.2. Figure 7.3 demonstrates the macrostructure of friction stir welded joints at various rotational velocities.

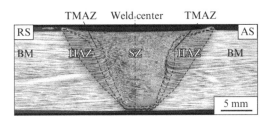

RS – Retreating Side; AS – Advancing Side; SZ – Stir Zone; HAZ – Heat Affected Zone; TMAZ –

Thermo Mechanically Affected Zone; BM – Base Metal

**FIGURE 7.2**
Transverse macrostructure of the weld of the steel plate.

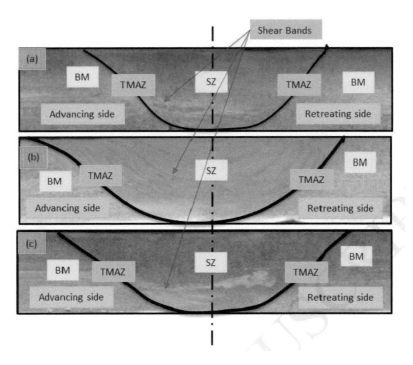

**FIGURE 7.3**
Macrostructure of FS welded joints: (a) 400, (b) 600, and (c) 800 rev./min [21]. (With kind permission from Elsevier.)

## 7.5 Tool Materials

The necessities for an FSW tool in HTM are critical. The tool must have the adequate quality to compel the weld material at softening temperatures more than 1000°C (1830°F). The tool should likewise be impervious to weariness, break, mechanical wear, and synthetic responses with the air and the welded material. Nowadays, there exist mainly two classes of materials that satisfy such prerequisites: refractory metals and super-abrasive tools.

### 7.5.1 Refractory Metal Tools

The first class of tool materials to be used for FSW of HTM is refractory metal tools. Initially, the tool materials were considered proprietary. Eventually, however, the composition of the tools was revealed. The different sorts of tools utilized for welding maraging which demonstrates the degree of tool wear are shown in Table 7.1. Tungsten was used as a tool material in a significant number of the initial welds executed. Tungsten seemed to have the adequate hot quality to fill in as an FSW device. This required preheating of

**TABLE 7.1**

Different Tools Used for Welding Maraging Steel Showing the Extent of Tool Wear [22]

| S. No. | Type of Tool | Final Shape of Tool | Macro View |
|---|---|---|---|
| 1. | WC | 20mm | Tungsten inclusion / weld / 800μm |
| 2. | WFe | 20mm / 5.4mm | Tungsten rich inclusion / weld / 800 μm |
| 3. | WMo | 20mm / 7mm | 25 μm |

the device to temperatures over 300°C (570°F) and boring of a pilot opening for the device. Subsequent device materials involved increments of up to 25% Re to tungsten, which brought down the changing temperature beneath that of room temperature.

Tungsten–rhenium tools demonstrate expanded break opposition and enhanced wear obstruction when compared with pure tungsten, and seem to have turned into the most generally utilized recalcitrant metal. Advancement of creation forms keeps on improving the device life of tungsten–rhenium tools. Molybdenum was utilized on no less than one event as a device material for FSW of steel. Early tungsten and tungsten–rhenium tools demonstrated an inclination to wear quickly in the weld, prompting naturally visible incorporations of tool material in the weld zone. Subsequent tools turned out to be considerably more impervious to this issue; however, the device material regularly keeps on dissolving in the weld, leaving a tungsten-improved mix zone. Unmanageable metal tools are utilized to weld low-carbon steels, ferrite stainless steels, carbon–manganese steels, austenitic stainless steels, etc. Tungsten–rhenium tools indicate excellent crack sturdiness, and so they may be utilized in generally thick welds (up to 13 mm in a solitary pass). Reported tool life ranges from one-quarter of a meter to approximately 4 m.

### 7.5.2 Super-Abrasive Tools

Super-abrasives are materials that are shaped in presses under outrageous temperature and weight. Polycrystalline diamond (PCD) and PCBN are the two super-abrasives that are used as a part of FSW. These two super-abrasives comprise little gems of ultra-hard material (diamond or CBN) reinforced together in a skeletal lattice with a moment stage material that fills in as an impetus for the development of the grid.

PCD has been used for aluminum–lattice alloys fortified with boron carbide besides particulate silicon carbide or alumina. This additionally guarantees as an effective tool material for welding titanium, in spite of the fact that this work is just in a preparatory stage.

It should be noted that polycrystalline cubic boron nitride is mainly used to weld carbon steels, carbon–manganese steels, high-quality, low-compound (HSLA) steels, high-quality pipeline steels, austenitic stainless steels, duplex stainless steels, double-stage steels, nickel-base alloys, and other fascinating combinations. It has also been tried in titanium compounds with conflicting outcomes. On occasion, it performs well; at others, compound responses with the workpiece results in quick wear.

Generally, the super-abrasive supplies are found just in little pieces because of the high weight required for assembling. Moreover, these materials are extremely troublesome or difficult to braze. Trials of PCBN tools in 316L stainless steel demonstrated device life of 1–4 m (3–13 ft), with life restricted by the break. Continuous endeavors at enhancing the plan of the alloy tools, with their upgrades in the review of the PCBN, have incredibly lessened the propensity of the device to break and have essentially expanded its life. The latest tool life test carried out on PCBN tools demonstrated a device life of 80 min in 1018 steel.

Polycrystalline cubic boron nitride tools create an uncommonly smooth surface on the weld. This resulting smoothness is believed to be due to the low coefficient of grinding among PCBN and the welding metal. The significant confinement in PCBN devices is the most extreme profundity of the weld. Even though a pin 13 mm (0.5 in.) long has been tried, for down-to-earth purposes, the most extreme profundity of welding right now is 10 mm (0.4 in.). Continuous endeavors in the plan of PCBN tools have to prompt increments in pin length. Uber Stir Technologies, the supplier of PCBN tools, does have plans to accomplish a 13-mm weld profundity within one year.

In recent years, critical endeavors used in creating harder, more wear-safe evaluations of PCBN transpired. Accomplishments to understand the impact of various fastener stages, the proportion of CBN, and grain measure circulations of CBN on execution were explored. The execution got assessed using a turning test on 304L stainless steel. Those evaluations displaying more prominent wear obstruction in the turning tests were assessed employing FSW in 304L. The PCBN review improvement program turned out to be very effective harder, and further wear-safe evaluations of PCBN were produced. Notwithstanding enhanced wear obstruction, the enhanced durability of the new evaluations has empowered both more profound weld entrance (up to 12 mm, or 0.47 in.) and strung compose highlights to be fused into the tool outline.

## 7.6 Tool Wear

The photographic pictures and optical macrographs of the FS welding joints created utilizing FSW devices, T1 and T2 are illustrated in Figures 7.4 and 7.5, separately. As evident from the Figure, the joints are devoid of outside or inward large-scale imperfections, for example, funneling, splits, and voids.

Referring to Figure 7.5, the half C curves successively formed on the top surface of the joint for tool T1 are more distinct than that for the joint made using tool T2. Also, for the joint made using tool T2, the weld surface was observed with a mild bluish color, probably due to the distribution of tool

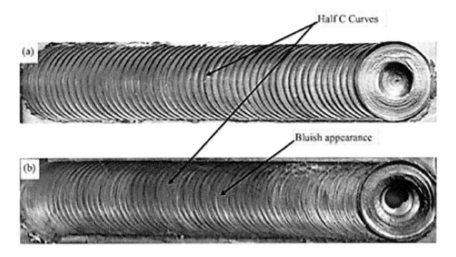

**FIGURE 7.4**
Images of FS welded 316L stainless steel joints produced by (a) tool T1 and (b) tool T2 [19]. (With kind permission from Elsevier.)

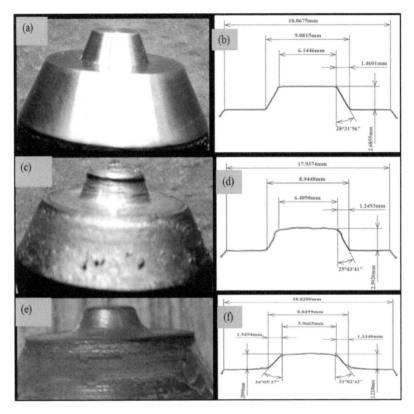

**FIGURE 7.5**
Tool pin and shoulder profiles before and after welds: (a and b) before welding; (c and d) after welding for tool T1; and (e and f) after welding for tool T2 [19]. (With kind permission from Elsevier.)

wear debris in the weld zone. The studies show that the tool material plays a significant role in the performance of the joints as the tool wear and tool deformation affect the microstructural and mechanical characteristics of the joints and the tungsten lanthanum oxide tool is found superior over the tungsten heavy alloy tool [20].

## 7.7 Equipment Requirements for FSW of Steels

It should be noted that the essential distinction concerning FSW of delicate amalgams (Al, Cu, Mg) and high softening temperature alloys (steels, stainless steels, nickel combinations) is the high temperature required for welding. Temperatures, are in excess of 800°C when FSW of higher unstiffening temperature combinations is carried out. The essential tool apprehension is a transference of heat through the tool into other regions like the shaft. This problem is ordinarily alleviated by the utilization of certain type of fluid cooling in the axle or the tool holder. Machine hardness is a necessity, when particularly PCBN category of devices are used. The PCBN category of tool material is fragile. Inadequate firmness in the casing on which the axle is mounted or the ways and rails on which the welding table is fixed can make the device "surf" amid dive and welding. Surfing can make expansive dynamic loads on the tool, which can cause pointless harm and untimely disappointment of hard tooling like PCBN; however, it can likewise be a downside for another tooling also. Moreover, inadequate firmness can make a misalignment between the tool and the planned welding joint [13,21].

## 7.8 Microhardness Evolution

As the microstructures fluctuate inside various zones of the FSW joint, hardness tests are utilized to portray the joint and give signs of the particular areas. The rigidity conveyances on the transverse cross-area of joints being welded in various turn speeds are plotted in Figure 7.6. The firmness bends

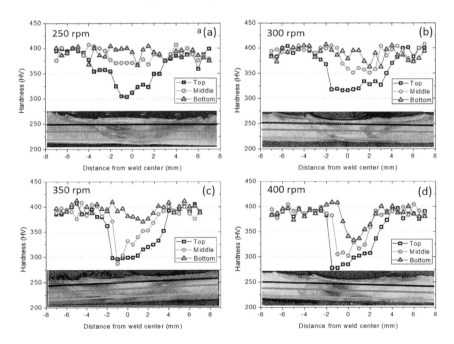

**FIGURE 7.6**
Hardness distribution in the joint welded at different rotational speeds: (a) 250, (b) 300, (c) 350, and (d) 400 rpm [9]. (With kind permission from Elsevier.)

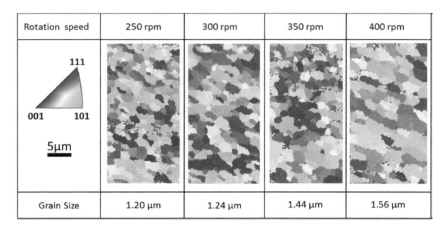

| Rotation speed | 250 rpm | 300 rpm | 350 rpm | 400 rpm |
|---|---|---|---|---|
| 111 ▲ 001 101  5μm | | | | |
| Grain Size | 1.20 μm | 1.24 μm | 1.44 μm | 1.56 μm |

**FIGURE 7.7**
Grain morphology of the top stir zone [9]. (With kind permission from Elsevier.)

show a valley shape that is described by a diminishing of rigidity in the SZ. These valleys do not appear to be symmetric, particularly at the upper turn speeds [9,14]. Because of the vast mis-sharpening amid the created procedure, the grains in the base metal (BM) get intensely prolonged. The constituent substances can assuage the forced vitality amid FSW and create new grains through crystallization in the SZ. Because the grain morphology gets altered in the recrystallization in the course of the FSW, the properties of the joint can be influenced. Grain morphology of the best blend zone and mix zone focus is illustrated in Figure 7.7. Expanding the speed of rotation from 250 to 400 rpm brings about cruder CDRX grains in the best SZ. The grain size of the SZ focus has a convoluted connection with the speed of rotation that is dictated by the interaction amid the temperature and the twisting strain [9,15–19].

## 7.9 Effect of Process Parameters

Lower welding speed resulted in higher temperature, and a slower cooling rate in the welding zone causes grain growth, which ensued in low tensile strength and toughness of impact. The joints made with low device rotational speed brought about the nonattendance of stage change and comprised of fine-grained ferrite structure [20]. The rigidity of the joints created at bringing down rotational speed is overmatched when compared with the BM. The tool shoulder width has a straightforwardly corresponding association with the heat generation because of contact in the form of friction. Figure 7.8 indicates the selection of a range of process parameters in FSW of AISI 409M Ferrite Stainless Steel (Table 7.2).

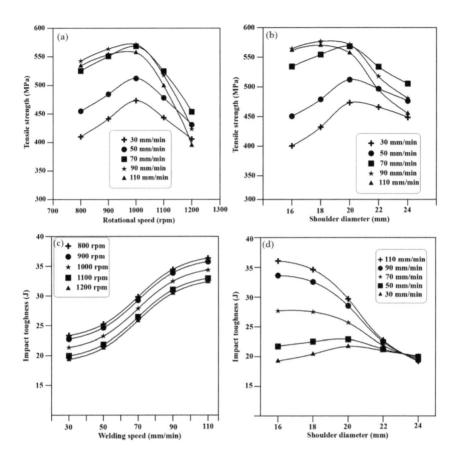

**FIGURE 7.8**
(a) Effect of rotational and welding speed on tensile strength. (b) Effect of shoulder diameter and welding speed on tensile strength. (c) Effect of welding speed and rotational speed on impact toughness. (d) Effect of shoulder diameter and welding speed on impact toughness [20]. (With kind permission from Springer.)

The sticking torque represents the resistance of the plasticized material against flow around the tool. An ideal measure of material flow around the device with least obstruction is required in order to achieve a decent weld and improved tool life in FSW [20].

## 7.10 Microstructural Evolution

The WN consists of three distinct regions, as shown in Figure 7.9, influenced by tool, shoulder, and their behavior inside the plate. The specked lines are the area extended because of the high heat input and cooling rate.

**TABLE 7.2**

Selection of a Range of Process Parameters in Friction Stir Welding of AISI 409M Ferrite Stainless Steel [20]

| S. No. | Parameter | Range | Macrostructure | Observation |
|---|---|---|---|---|
| 1. | Rotational speed | <800 rpm | | Tunnel defect—Insufficient material transportation due to lesser frictional heat |
| | | >1200 rpm | | Tunnel defect and flash formation—Excessive release of stirred materials to the upper surface, which resultantly left voids in the stir zone and formation of flash due to higher rotational speed |
| 2. | Welding speed | <30 mm/min | | Root pining due to excessive contact time between the hot weld metal and the steel backing plate. Also caused thermal damage to the tool |
| | | >110 mm/min | | Groove defect—Insufficient frictional heat hinders the viscoplastic material flow. Also, high flow stresses in the material can cause the tool to fail |
| 3. | Shoulder diameter | <16 mm | | Surface defect—The area between the shoulder diameter and pin diameter is reduced which in turn reduced the material flow from the leading to trailing edge |
| | | >24 mm | | Tunnel defect—Excessive heat generation and inadequate material flow and mixing |

The microstructure in the weld piece at typical cooling demonstrates ferrite and pearlite; however, when the cooling rate increases, weld chunk indicates martensite. HAZ zone 1 demonstrates a polygonal ferrite structure, whereas HAZ zone 2 indicates greater firmness than HAZ zone 1 comprising tempered

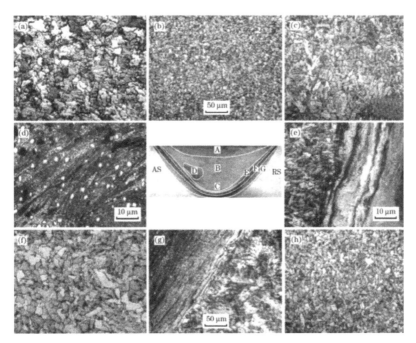

**FIGURE 7.9**
Optical micrographs of various regions of friction stir-welded mild steel [27]. (With kind permission from Elsevier.)

martensite. The TMAZ zone is not seen in the welded structure. In FSW of 304L and 316L steels, fractional recrystallization in the TMAZ is watched. In the HAZ zone of HSLA-65 steel weld, Widmanstatten ferrite and ferrite carbide totals were observed in the blend zone [27]. While in DH-36 steel, bainite and martensite structures were seen in the stir zone and HAZ zone little measures of pearlite are dispersed. FSW weld of C-Mn steel demonstrates TMAZ zone with fine-grained structures close to the mixing zone. The BM comprises of allotriomorphic ferrite and pearlite. Equiaxed earlier austenite grain structures were found in the TMAZ. For high-carbon steel at A3 temperature, the ferrite structure and globular cementite structure were changed entirely to an austenite structure amid FSW, and in the wake of welding the material experienced strong state change from austenite structure to martensite structure; however, in HAZ zone, martensite is not seen [22–25].

At the point when an external heat is supplied by the gas burner, the carbon atoms diffuse amid the cooling and lead to the development of martensite. Disfigurement in HAZ may build the measure of austenite grain per unit volume. This would prompt a relating huge increment in the grain limit nucleation rate of allotriomorphic ferrite. Warmth input gave the best relationship post-grinding blend weld microstructures. Warmth input shows a direct association with ferrite grain measure and bainite slat estimate. Ferrite grain measure, as well as bainite strip estimate, expanded 150% with an expansion in the warm contribution of 2.27 kJ/mm [4,26–30].

The quality of the weld is tried for malleability and effect, and the rigidity was found to be more than the BM. FSW is accomplished on M190 steel; the rate of warmth info and cooling rate amid welding is connected from the beforehand accessible observational connections. The mechanical properties of the joints can be understood from the miniaturized scale and full-scale auxiliary investigation. From the smaller scale basic perception it is discovered that the TMAZ zone found in aluminum combinations isn't seen in steels. The weld zone was established to surpass a pinnacle temperature of 1000°C, and in the HAZ zone, fine pearlites are arbitrarily circulated in the equiaxed ferrite grains. At the point when the plate thickness of steel is under 6 mm a single passweld is effective, yet when the plate thickness is more than 6 mm, two passes are important to create a productive weld [26,31–46].

## 7.11 Defects in FSW

In spite of numerous focal points related to the FSW procedure, the system does not generally deliver deformity free joints. Controlling the FSW procedure keeping in mind the end goal to create fantastic weld joints is a test because of the sum of considerations related to the FSW procedure. Such considerations incorporate free (for example, tool rotational/navigate rates) and ward (for example, powers and torque) welding process constraints, device material, tool plan, workpiece material, and its thickness. The following are the different kinds of deformities [16,47]:

- Wormholes, voids, and passages in the base of the weld joints, presumably because of inadequate warmth input and the need in material flow.
- Kissing Bonds Cracks, however, in close contact normally situated at the weld root; they are materials in the absence of synthetic and mechanical holding.
- Root staying is caused by intemperate warmth and contact time bringing about the workpiece adhering to the backing plate.
- Streak development and material diminishing: caused basically by over the top warmth because of unreasonable pivotal powers.
- Weld Root Flaw Cracks beginning with the base of the workpiece at irregular planes toward the welded zone.
- The oxidation process came about because of higher temperatures without any gas shield amid the FSW procedure.

## 7.12 Conclusions

Although FSW is a widespread interest for most of the upcoming researchers in the area of welding, it finds its importance in welding high-strength steels. As a special interest to review the recent developments in FSW of steel welding, this chapter summarized the basic concepts to understand the formation of the weld and its process parameter which functions to give such a permanent joint. It can be seen how the tool design and welding process parameter contribute to the formation of a joint between similar and dissimilar welds. Butt and lap joints are the two widely used types of joints. Most of the welds are performed on butt type of joints rather than lap joint because of elimination of surface preparation, easy to fix, clamp, and perform welding. In most of the cases, the FSW welding is done with a cylindrical threaded pin, but there are very few applications in the tapered pin. The tapered threaded pin can give better strength and uniform weld throughout the length of the plate as discussed in tool design. Vertical mixing at high velocity and tool wear can be minimized through this design. Rotating speed, traverse speed, and tilt angle are the majors factors that were found to influence the weld formation more than the other process parameter such as plunge depth and force exerted by the tool compared to weld, where the tool rotates perpendicular to the workpiece, when placed at an angle facing the direction gave high-strength joint.

## References

1. R. Ramesh, I. Dinaharan, R. Kumar and E.T. Akinlabi. "Microstructure and mechanical characterization of friction stir welded high strength low alloy steels". *Materials Science and Engineering: A* 687, 2017, 39–46.
2. R. Ramesh, I. Dinaharan, E.T. Akinlabi and N. Murugan. "Microstructure and mechanical characterization of friction-stir-welded dual-phase brass". *Journal of Materials Engineering and Performance* 27, 2018, 1544–1554.
3. A. Ozekcin, H.W. Jin, J.Y. Koo, N.V. Bangeru, R. Ayer, G. Vaughn, R. Steel and S. Packer. "A microstructural study of friction stir welded joints of carbon steels". *ISOPE*, Toulon, 2004.
4. C.J. Sterling, T.W. Nelson, C.D. Sorensen, R.J. Steel and S.M. Packer. "Friction stir welding of quenched and tempered C–Mn steel". In R.S. Mishra and Z.Y. Ma (eds.). *Friction Stir Welding and Processing II*. Materials Park, OH: The Minerals, Metals, and Materials Society, 2003, 165–171.
5. R.J. Steel, C.D. Sorensen, C. Pettersson, Y.S. Sato, T.W. Nelson, C.J. Sterling and S.M. Packer. "Friction stir welding of SAF 2507 (UNS S32750) super duplex stainless steel". *Stainless Steel World 2003*, Maastricht, November 11–13, 2013.
6. H. Sarlak, M. Atapour and M. Esmailzadeh. "Corrosion behaviour of friction stir welded lean duplex stainless steel". *Materials & Design (1980–2015)* 66, Part A, 2015, 209–216.

7. A. Toumpis, A. Galloway, S. Carter and N. McPherson. "Development of a process envelope for friction stir welding of DH36 steel—A step change". *Materials & Design (1980–2015)* 62, 2014, 64–75.

8. A.J. Ramirez, T.F.C. Hermenegildo, V.F. Pereira1, J.A. Avila1, T.F.A. Santos, P.R. Mei, L.P. Carvalho, R.R. Marinho and M.T.P. Paes. "Friction stir welding of steels for the oil and gas industry". *Proceedings of the 1st International Joint Symposium on Joining and Welding*, Osaka, November 6–8, 2013, 75–79.

9. W. Han, P. Liu, X. Yi, Q. Zhan, F. Wan, K. Yabuuchi, H. Serizawa and A. Kimura. "Impact of friction stir welding on recrystallization of oxide dispersion strengthened ferritic steel". *Journal of Materials Science and Technology* 34, 2018, 209–213.

10. M. Matsushita, Y. Kitani, R. Ikeda, K. Oi and H. Fujii. "Development of friction stir welding of high strength steel sheet". *Proceedings of the 1st International Joint Symposium on Joining and Welding*, Osaka, November 6–8, 2013, 87–93.

11. Y. Miyano, H. Fujii, Y. Sun, Y. Katada, S. Kuroda and O. Kamiya. "Mechanical properties of friction stir butt welds of high nitrogen-containing austenitic stainless steel". *Materials Science and Engineering: A* 528, 2011, 2917–2921.

12. S. Wu, Y. Morisada, R. Ueji, H. Fujii, Y.F. Sun and C. Shiga. "Minor metal reduction of high tensile strength steel by friction stir welding". *Proceedings of the 1st International Joint Symposium on Joining and Welding*, Osaka, November 6–8, 2013, 115–117.

13. F.C. Liu, Y. Hovanski, M.P. Miles, C.D. Sorensen and T.W. Nelson. "A review of friction stir welding of steels: Tool, material flow, microstructure, and properties". *Journal of Materials Science and Technology* 34, 2018, 39–57.

14. N.R. Mandal. *"Ship Construction and Welding"*. *Springer Series on Naval Architecture, Marine Engineering, Ship Building and Shipping* 2, 2016, 91–114.

15. Russell Steel. "Friction stir welding: New developments for oil and gas industry". *Proceedings of International Oil and Gas Conference and Exhibition*, IOGCEC, 2010.

16. M. Al-Moussawi and A.J. Smith. "Defects in friction stir welding of steel". *Metallography, Microstructure, and Analysis* 7, 2018, 194–202.

17. V.V. Sagaradze, V.I. Shalaev, V.L. Arbuzov, B.N. Goshchitskii, Y. Tian, W. Qun and S. Jiguang. "Radiation resistance and thermal creep of ODS ferritic steels". *Journal of Nuclear Materials* 295, 2001, 265–272.

18. M. Matsushita, Y. Kitani, R. Ikeda, M. Ono, H. Fujii and Y.-D. Chung. "Development of friction stir welding of high strength steel sheet". *Science and Technology of Welding and Joining*, 16, 2011, 181–187.

19. S. Shashi Kumar, N. Murugan and K.K. Ramachandran. "Influence of tool material on mechanical and microstructural properties of friction stir welded 316L austenitic stainless steel butt joints". *International Journal of Refractory Metals and Hard Materials* 58, 2016, 196–205.

20. A.K. Lakshminarayanan and V. Balasubramanian. "Understanding the parameters controlling friction stir welding of AISI 409M ferritic stainless steel". *Metals and Materials International* 17, 2011, 969–981.

21. S. Shashi Kumar, N. Murugan and K.K. Ramachandran. "Microstructure and mechanical properties of friction stir welded AISI 316L austenitic stainless steel joints". *Journal of Materials Processing Technology* 254, 2018, 79–90.

22. S.D. Meshram, G. Madhusudhan Reddy and S. Pandey. "Friction stir welding of maraging steel (Grade-250)". *Materials and Design* 49, 2013, 58–64.

23. R. Ueji, H. Fujii, L. Cui, A. Nishioka, K. Kunishige and K. Nogi. "Friction stir welding of ultrafine grained plain low carbon steel formed by the martensite process". *Materials Science and Engineering: A* 423, 2006, 324–330.

24. A. Ozekcin, H.W. Jin, J.Y. Koo, N.V. Bangaru, R. Ayer, G. Vaughn, et al. "A microstructural study of friction stir welded joints of carbon steels". *International Journal Offshore Polar Engineering* 14, 2004, 284–288.

25. D.-H. Choi, et al. "Hybrid friction stir welding of high-carbon steel". *Journal of Materials Science and Technology* 27(2), 2011, 127–130.

26. H. Fujii, L. Cui, N. Tsuji, M. Maeda, K. Nakata and K. Nogi. "Friction stir welding of carbon steels". *Materials Science and Engineering: A* 429, 2006, 50–57.

27. A.K. Lakshminarayanan, V. Balasubramanian and M. Salahuddin. "Microstructure tensile and impact toughness properties of friction stir welded mild steel". *Journal of Iron and Steel Research, International* 17(10), 2010, 68–74.

28. M. Ghosh, K. Kumar and R.S. Mishra. "Friction stir lap welded advanced high strength steels: Microstructure and mechanical properties". *Materials Science and Engineering: A* 528, 2011, 8111–8119.

29. L.Y. Wei and T.W. Nelson. "Correlation of microstructures and process variables in FSW HSLA-65 steel". *Welding Journal* 90, 2011, 95s–101s.

30. M. Posada, J.J. DeLoach, A.P. Reynolds, R.W. Fonda and J.P. Halpin. "Evaluation of friction stir welded HSLA-65". *Proceedings of the Fourth International Friction Stir Welding Symposium*, Abington, UK: TWI, 2003.

31. P. Sinha, S. Muthukumaran and S.K. Mukherjee. "Analysis of first mode of metal transfer in friction stir welded plates by image processing technique". *Journal of Materials Processing Technology* 197, 2008, 17–21.

32. S. Muthukumaran, K. Pallav, V.K. Pandey and S.K. Mukherjee. "A study on the electromagnetic property during friction stir weld failure". *International Journal of Advanced Manufacturing Technology* 36, 2008, 249–253.

33. M.M. Attallah and H.G. Salem. "Friction stir welding parameters: A tool for controlling abnormal grain growth during subsequent heat treatment". *Materials Science and Engineering: A* 391, 2005, 51–59.

34. W.M. Thomas, K.I. Johnson and C.S. Wiesner. "Friction stir welding-recent developments in tool and process technologies". *Advances Engineering Materials* 5, 2003, 485–490.

35. S. Muthukumaran and S.K. Mukherjee. "Multi-layered metal flow and formation of onion rings in friction stir welds". *International Journal of Advanced Manufacturing Technology* 38, 2008, 68–73.

36. M.W. Mahoney, C.G. Rhodes, J.G. Flintoff, R.A. Spurling and W.H. Bingel. "Properties of friction-stir-welded 7075 T651 aluminium". *Metallurgical and Materials Transaction A* 29, 1998, 1955–1964.

37. R. Nandan, G.G. Roy, T.J. Lienert and T. DebRoy. "Numerical modelling of 3D plastic flow and heat transfer during friction stir welding of stainless steel". *Science and Technology of Welding and Joining* 11, 2006, 526–537.

38. R. Nandan, G.G. Roy, T.J. Lienert and T. DebRoy. "Three-dimensional heat and material flow during friction stir welding, of mild steel". *Acta Materialia* 55, 2007, 883–895.

39. P.A. Colegrove and H.R. Shercliff. "Development of Trivex friction stir welding tool. Part 1—Two-dimensional flow modelling and experimental validation". *Science and Technology of Welding and Joining* 9, 2004, 345–351.

40. P.A. Colegrove and H.R. Shercliff. "Development of Trivex friction stir welding tool. Part 2—Three-dimensional flow modelling". *Science and Technology of Welding and Joining* 9, 2004, 352–361.

41. M. Guerra, C. Schmidt, J.C. McClure, L.E. Murr and A.C. Nunes. "Flow patterns during friction stir welding." *Materials Characterization* 49, 2002, 95–101.

42. P. Sinha, S. Muthukumaran, R. Sivakumar and S.K. Mukherjee. "Condition monitoring of the first mode of metal transfer in friction stir welding by image processing techniques". *International Journal of Advanced Manufacturing Technology* 36, 2008, 484–489.

43. J.A. Schneider and A.C. Nunes. "Characterization of plastic flow and resulting microtextures in a friction stir welding." *Metallurgical and Materials Transactions B* 35, 2004, 777–783.

44. K.N. Krishnan. "On the formation of onion rings in friction stir welds". *Materials Science and Engineering: A* 327, 2002, 246–251.

45. G.G. Roy, R. Nandan and T. DebRoy. "Dimensionless correlation to estimate peak temperature during friction stir welding". *Science and Technology of Welding and Joining* 11, 2006, 606–608.

46. S. Muthukumaran and S.K. Mukherjee. "Two modes of metal flow phenomenon in the friction stir welding process". *Science and Technology of Welding and Joining* 11, 2006, 337–340.

47. R. Nandan. "Recent advances in friction-stir welding—Process, weldment structure and properties". *Progress in Materials Science* 53, 2008, 980–1023.

# 8

## Friction Stir Spot Welding of Similar and Dissimilar Nonferrous Alloys

**R. Palanivel, I. Dinaharan, and R. F. Laubscher**
*University of Johannesburg*

**CONTENTS**

8.1 Overview ................................................................................................. 159
8.2 Joining Mechanism ............................................................................... 161
8.3 Microstructural Evolution .................................................................. 161
    8.3.1 Material Flow ............................................................................ 162
    8.3.2 Hook Formation ....................................................................... 163
    8.3.3 Local Melting ........................................................................... 165
    8.3.4 Cracking .................................................................................... 165
    8.3.5 Intermetallic Formation ......................................................... 166
8.4 Effects of Process Parameter .............................................................. 168
8.5 Effect of Tool Design ............................................................................ 172
8.6 Process Variants .................................................................................... 176
    8.6.1 Refilling Spot Welding ........................................................... 176
    8.6.2 Double-Side Spot Welding ..................................................... 177
    8.6.3 Pin-Less Tool ............................................................................ 178
    8.6.4 External Heating ...................................................................... 178
    8.6.5 Particle Addition ..................................................................... 180
8.7 Future Trends ........................................................................................ 181
    8.7.1 Ferrous Alloys .......................................................................... 181
    8.7.2 Nonmetals ................................................................................. 183
8.8 Summary ................................................................................................. 185
References ......................................................................................................... 185

## 8.1 Overview

Resistance spot welding (RSW) is generally employed to join metal sheets in the automobile industries. This RSW may regularly be exposed to intricate multiaxial loads amid any kind of service. RSW makes common splits or scores along the weld circuits due to its exceptional nature. This drawback required the advancement of stress intensity factor design techniques to account for the main and curved crack fronts at critical locations [1]. Friction stir welding (FSW) was, for the first time, created by the The Welding Institute (TWI) in the year 1991 to join sheet metal composites. This innovation exhibits critical advantages including less contortion, bring down lingering stresses while holding the mechanical properties of the parent metal [2,3]. In 2001, friction stir spot welding (FSSW) was produced and begun to replace RSW in some of selected applications in the automobile industries to a great extent to evade its related drawbacks [4]. The most important difference between RSW and FSSW was that no softening of the parent metal occurs in the latter. This suggests a noteworthy preferred standpoint as the weld joint is shaped without the combination of the base metals (BMs), which maintains a strategic distance from the development of specific kinds of defects in the welding (for example, pores and splits) in the weld creases. Hence, this welding system is prominently appropriate for metal joining without the related deformities because of the melting that takes place [5]. Some of the typical advantages of FSSW are as follows:

- The only energy required for FSSW is the electricity needed to rotate and drive the tool. This is typically appreciably less than required for fusion welding.
- No weld consumables are used. FSSW tools typically have an extended lifespan.
- No unsafe emanations and condition well disposed. No weld splash, clamor, and decreased vapor discharge amid FSSW.
- Little welding deformation. FSSW is a solid-state welding process without melting of materials, so distortions are smaller than for RSW.
- High repeatability and consistency due to its simple joining mechanism and relatively few significant process parameters.
- No or limited weld preparation is required. No need for extensive surface cleaning, drilling, riveting, or bolting.

The process parameters that affect the weld quality in terms of strength, microstructural properties, material flow, and heat generation are rotational speed, dwell period, and plunge depth. Significant tool geometry parameters that also affect weld quality incorporate shoulder distance across the shape of the pin, profile of the pin, and its length [6–9]. Some of the typical features of the friction stir spot welds are higher strength, better fatigue life, lower distortion, less residual stress, and better corrosion resistance. High joint strengths can be achieved without porosity, cracks, and contamination [10–12]. This chapter introduces the current state of FSSW technology along with recent developments related to the joining mechanism of microstructure development and its mechanical properties. The impact of process parameters on the weld quality is also addressed in detail.

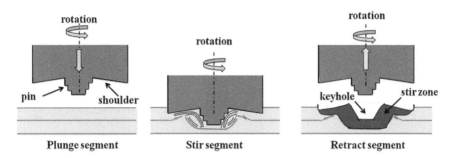

**FIGURE 8.1**
Schematic representation of FSSW mechanism [16]. (With kind permission from Elsevier.)

## 8.2 Joining Mechanism

FSSW occurs mainly in three stages: plunging, stirring, and retraction (see Figure 8.1). The tools used for FSSW typically consist of two parts: a pin or probe, and a shoulder. The pin is intended to upset the faying surfaces of the workpieces, disrupt the faying surfaces of the workpieces, shear and transport the material around it, and produce heat due to deformation and friction. The tool shoulder produces the majority of frictional heat that is conducted from the surface toward the subsurface regions of the workpieces. The shoulder also constrains the flow of plasticized material and is responsible for the downward forging action. Commonly amid FSSW, an extraordinarily planned pivoting round and hollow apparatus (with shifting end geometry and test stick) is at first dove into the upper sheet. The pivoting apparatus contacts the upper sheet with a connected descending power. This power is upheld by an instrument arranged underneath the lower sheet. The descending power and the rotational speed are kept up for a fitting time to produce adequate frictional warmth. The warmed and along these lines diminished material contiguous to the apparatus at that point disfigures plastically, and a solid-state bond happens between the surfaces of the upper and lower sheets. Finally, the instrument is withdrawn leaving a trademark opening amid the joint due to the pin [6,7].

## 8.3 Microstructural Evolution

The heat as well as plastic flow occurring due to the rotation of the tool results in the unique microstructural properties and mechanical characteristics associated with the joint. Figure 8.2 presents the typical microstructural zones associated with an FSSW joint (AA 6061-T4 sheets).

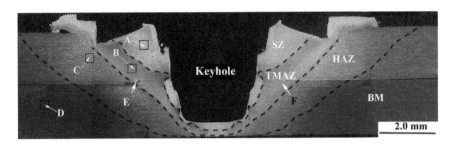

**FIGURE 8.2**
Schematic illustrations of the different zones after FSSW. (SZ: stir zone, TMAZ: thermomechanically affected zone; and HAZ: heat-affected zone) [13]. (With kind permission from Elsevier.)

Normally, the microstructure advances as an element of the spiral separation from the primary rotational hub. At the outer layer of the joint (farthest from the primary pivot), no alteration of the microstructure of the BM happens. The heat-affected zone (HAZ) is the first region radially inwards from the unaffected BM where microstructural modification occurs due to the thermal load. The next region includes materials that have been modified due to the thermomechanical action of the revolving tool. The thermomechanical affected zone (TMAZ) includes significant plastic deformation due to the tool stirring action and typically leads to an escalation of the average grain size. As a final point, the last region at the center is referred to as the weld nugget. Recrystallization of the original grains typically occurs in this region with a fine (few micrometers) uniform grain structure being the result [6,7].

### 8.3.1 Material Flow

Material flow amid the FSSW is regularly explored utilizing tracer material strategies. Figure 8.3 presents a model for material flow. During the initial stage where the pin penetrates the upper surface (before the shoulder contact), the stick, for the most part, expels the material downwards, thereby triggering the connecting lower sheet material to move upwards into the upper sheet. After the pivoting shoulder comes into close contact with the upper sheet, three unmistakable regions are created in the weld [14,15]. They are the stream progress zone promptly underneath the apparatus bear, the mixing zone around the stick, and the torsion zone underneath the stick that advances because of the mix of rotational, even, and vertical movements of the plasticized material. Figure 8.3 also includes a recent model put forward to describe the material flow during FSSW based on experimental observations [16,17]. In the stream change zone, material stream is instigated by the collaboration of the turning shoulder and stick. Because of a rotational symmetry, every material unit spirals along the surface of a virtual rearranged cone to achieve its concurrent circumferential

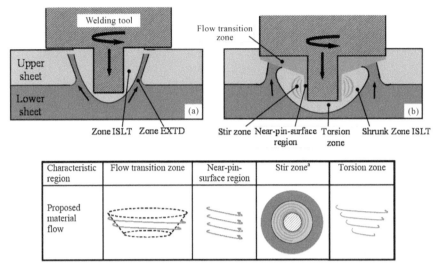

<sup>a</sup>A horizontal section shows the outward-spinning motion within the stir zone.

**FIGURE 8.3**
Schematic illustration of a new model of material flow: (a) before the shoulder contacts the upper sheet and (b) after the shoulder penetrates into the upper sheet [14]. (With kind permission from Elsevier.)

and downwards inclining movements. Joining of the upper and lower sheet materials happens, and the intermixed materials in this way stream toward the stick. Material stream in the mixing zone is commanded by the revolution of the stick. The material discharged from the stick surface through an outward-turning movement gives an inherent main thrust to the descending winding movement of the plasticized material along the stick. The blend zone is framed because of the individual commitments of the three material transport forms. These include (1) the upward movement of the lower sheet material and the incorporation of the upper and lower sheet materials, (2) the descending winding movement of the coordinated materials along the stick, and (3) the arrival of the incorporated material from the stick. As the blend zone grows, the material instantly outside the mixing zone streams toward the stick by following the mix zone limit and is constrained straightforwardly into the mixing zone. In the torsion zone, the material underneath the stick moves downwards in a twirling movement because of the impact of higher pressure torsion [16].

## 8.3.2 Hook Formation

A specific characteristic of a welded joint by friction stir spot method in lap configuration is the creation of a geometrical fault happening at the crossing point of the two welded sheets, referred to very often as a "hook" [18]. The basic geometry of a hook formation is depicted in Figure 8.4a. The

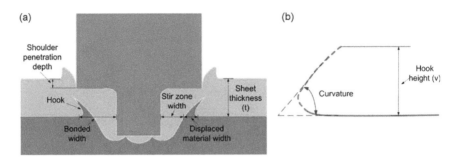

**FIGURE 8.4**
(a) Schematic representations showing how the stir zone and bonded width values, and shoulder penetration depths are measured in an AZ31 spot weld. (b) The method used to determine the height and the curvature of a hook region spot weld [19]. (With kind permission from Elsevier.)

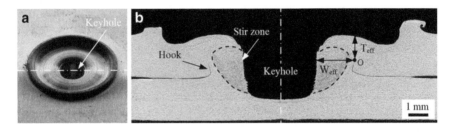

**FIGURE 8.5**
Macrographs of the friction stir spot welded joint (Vp, 60 mm/min and Vs, 6 mm/min): (a) surface appearance and (b) cross-sectional macrograph [18]. (With kind permission from Springer Nature.)

geometric parameter associated with a hook formation is presented in Figure 8.4b. Figure 8.5 presents a macro photo and sectional macrograph of an FSSW joint and the associated hook formation.

The incorporated material of the upper and lower sheets uproot downwards along the string to in the long run alter course to stream radially outwards when it achieves the stick tip. It at that point streams outwards and upwards before re-circling back toward the foundation of the stick and downwards once more. Since an expanding measure of the lower sheet material streams upwards and is incorporated with the upper sheet material, the mix zone increments in the estimate. The material present simply outside the blend zone is expelled by the mixing zone and uprooted upward. Therefore, the hook moves upwards and outwards from the hub of the keyhole alongside the material adjoining the blend zone. The snare region is dislodged upwards and outwards from the hub of the turning apparatus amid the stay time frame. The separation from the sheet crossing point to the tip of the snare area (the $(v/t)$ proportion) and the edge of the arch of the hook locale both increment notably amid the stay time frame [19].

### 8.3.3 Local Melting

It is, for the most part, expected that nearby softening and splitting does not happen in FSSW as the most noteworthy temperatures created inside the stir zone are not exactly those required for local liquefying [20,21]. The most astounding temperature accomplished in the stir zones of Al composite and Mg compound amid FSSW is eventually restricted by either the solidus temperature of the combination being referred to or the unconstrained liquefying temperatures of second-stage particles, which are contained in the "as gotten" BM. The warming rate amid the instrument entrance stage is controlled by the device rotational speed and ranges from 210 to 400°C.s$^{-1}$. The high warming rate amid rubbing during the spot welding of Al compound segments has exciting results. $\eta$, S, and T particles were formed in the Al7075-T6 base material, as they have inadequate time to liquefy and unexpectedly, hence the mix zone temperature achieves 475 ($\eta$), 480 (S) and 490 (T) °C. High apparatus rotational velocities prompt a reduction in the assessed strain rate due to the related unconstrained dissolving of second-stage particles ($\eta$, S, and T stages), which advances device slippage at the contact interface between the fringe of the turning instrument and adjoining material in the blend zone. High instrument rotational paces create a blend of high warming rate and high mix zone temperature that encourages neighborhood dissolving and device slippage [22,23]. The dissolved films normally edge, because of warming amid; (1) amid the device entrance organize in spot welding, (2) when the material is dislodged upwards amid device, penetration (3) when the instrument bear is in contact with the surface of the upper sheet. Figure 8.6 presents confirmation of eutectic film development in the mixing zone of an Al 7075 contact blend spot weld. The eutectic films are rich in Zn and Cu. The temperature in the mixing zone amid Al 7075 grinding blend spot welding surpasses the unconstrained softening temperatures of the second-stage particles 475 ($\eta$), 480 (S), and 490°C (T). The unconstrained liquefying of g particles ($\eta$) situated at grain limit regions creates liquefied eutectic films rich in Zn and Cu with the arrangement of an ($\alpha - $Al$+ $S$+ \eta$) ternary eutectic. The eutectic films seen in the mix zones of Al 7075 spot welds are regularly 50 μm long and 2 μm wide (see Figure 8.6). Towards the finish of the apparatus penetration arrangement of recrystallizedgrainshappens and second-stage particles, which were adjusted along grain limits, are incorporated into the blend zone.

### 8.3.4 Cracking

It is, for the most part, accepted that FSSW is free from a number of deformity arrangement issues ordinarily connected with fusion welding. This is anyway not generally the situation. As of late liquid penetration induced (LPI) splitting in the mix, the zone has been found in FSSW of AZ91 [24]. It is

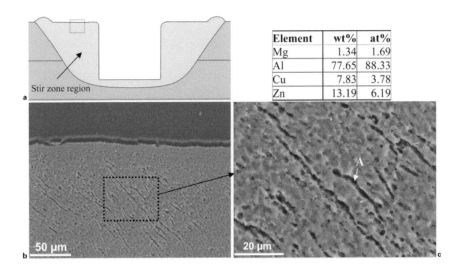

| Element | wt% | at% |
| --- | --- | --- |
| Mg | 1.34 | 1.69 |
| Al | 77.65 | 88.33 |
| Cu | 7.83 | 3.78 |
| Zn | 13.19 | 6.19 |

**FIGURE 8.6**

(a) Schematic of location in Al 7075 spot weld stir zone, (b) SEM image of melted films, and (c) detail of melted film and EDX chemical analysis at location A [22]. (With kind permission from Taylor & Francis.)

recommended that splitting amid AZ91, AZ31, and AM60 FSSW is dictated by the accompanying features [25,26]:

Figure 8.7 shows the arrangement of liquid eutectic films and splitting in the TMAZ area promptly adjoining the mix zone furthest point right on time in the stay time frame (dwell times of 0.05 and 0.1 s).

Coalescence of liquid eutectic films when the mix zone increses in width amid the dwell period. At the point when liquid eutectic films framed in the TMAZ area mixture, it gives the conditions to the entrance of a-Mg grain limits in the blend zone near its furthest point. A significant part of the examination, analyzing the fluid metal embrittlement has featured the split inception instead of the break proliferation arrange in disappointment since fluid film stage at grain limit areas, is amazingly quick. Thus, the blend of liquid eutectic films shaped in the TMAZ area encourages the fast entrance of a-Mg grain limits in the mixing zone near its furthest point. Tool extraction, in essence, does not advance splitting in the area. The hub development upwards to the pivoting apparatus, when the spot welding activity is ended, just isolates segments, which are as of now cracked. In AZ31 and AM60 spot, welds splitting are just obvious right on time in the stay time frame. The splitting inclination is lessened when the abide time increments past 2 s during and for AZ31.

## 8.3.5 Intermetallic Formation

The prospects of joining different amalgams are related to the development of weak intermetallic mixes (IMCs) in the mixing zone alongside the

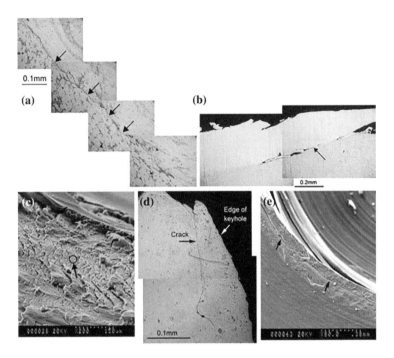

**FIGURE 8.7**
Cracking in the TMAZ region of AZ91, AM60, and AZ31 friction stir spot welds made using dwell times of 0.05 s (AZ31) and 0.1 s (AZ91 and AM60). (a) Cracking in the TMAZ of an AZ91 spot weld, see arrows. (b) Cracking in the TMAZ of an AM60 spot weld, see arrows. (c) Cracking surface in the TMAZ of an AM60 spot weld; the chemical composition at the location indicated on the keyhole surface is 79.0 wt% Mg, 20.9 wt% Al. (d) Cracking in the TMAZ of an AZ31 spot weld, see arrows. (e) Cracking immediately adjacent to the keyhole periphery in the TMAZ of an AZ31 spot weld (see arrows) [26]. (With kind permission from Springer Nature.)

geometrical highlights of the weld [26]. These IMCs are fragile and substantially harder when compared with the base material and go about as locales for simple break development. The thickness of the IMCs has an immaterial effect on the lap shear quality of the weld. However, the circulation of these IMCs does influence the weld quality [27–31]. Device turn rate, apparatus infiltration profundity, and stay time (which are on the whole generally in charge of the frictional warmth age) are some of the fundamental factors that impact the formation of IMCs. Choi et al. [30] studied the SEM micrographs and EDS results for the interface are as displayed in Figure 8.8. A response layer was seen at the interface between the Al and Mg amalgam plates amid FSSW. Regions A and E were considered as the Al and Mg substrates, respectively, suggesting that these areas were not influenced by dissemination. Considering the quantitative examination of consequences of these areas and the concoction synthesis of the BM, locales B and D were expected to be Al and Mg progress zones, suggesting that Mg and Al components were diffused into Al and Mg substrates, with the aftereffect of strong arrangement

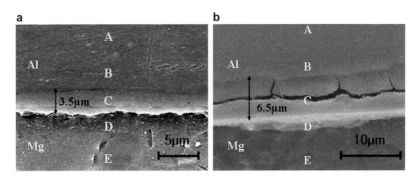

**FIGURE 8.8**
Microstructure and chemical compositions in the interface between Al and Mg alloy: (a) 1050 rpm–5 s, (b) 1600 rpm–30 s, and (c) result of spot analysis at the interface [30]. (With kind permission from Elsevier.)

development. Point C is the response layer and shows the presence of IMCs of Al and Mg. Figure 8.8 shows a response layer with the thickness ranging from 3.5 to 6.5 mm with an expanding instrument turn speed and stay time. Furthermore, a level break was seen in the thicker response layer, as shown in Figure 8.8b. As evident in the figure, the thickness of the IMC is firmly identified with the mechanical properties of the Al–Mg FSSW joint. These IMCs were considered to have been shaped by the dispersion behavior of Al and Mg iotas amid FSSW. The thickness of the IMC layer is an element of hardware turn speed and length. The IMC layer thickness increments with the rotational speed and device entrance profundity.

## 8.4 Effects of Process Parameter

Process parameters, for example, tool rotational speed, dwell time, feed rate, and plunge profundity, impact the mechanical and metallurgical properties

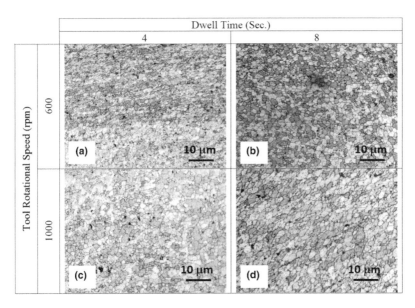

**FIGURE 8.9**
Microstructures of the SZ observed on AA5754 aluminum alloy sheets joined by FSSW using several tool rotational speeds and dwell times (1 mm from the periphery of the weld pinhole) [35]. (With kind permission from Springer Nature.)

of FSSW joints [32–43]. Run of themicrostructures of stir zones (SZs) observed for AA5754 aluminum compound sheets joined by FSSW utilizing a reduced rotational speeds and times are as shown in Figure 8.9. The SZ displayed a fine recrystallized equiaxed grain structure. Expanding the device rotational speed as well as the abide time builds the grain estimate in the SZ. This might be ascribed to an ascent in temperature because of the expanding apparatus rotational speed and additionally the abide time (expanded grating), which prompts grain development. Welds made with an apparatus rotational speed of 600 rpm and an abide time of 2 s showed the smallest grain measure at the SZ. Welding made with an apparatus rotational speed of 1200 rpm and a stay time of 10 s displayed the largest grain measure in the SZ [35].

The variety of a definitive force of tensile shear strength with device rotational speed at various stay times is depicted in Figure 8.10. The outcomes of the study uncovered that at consistent dwell time and increased tool rotational speed to a specific level increases the tensile shear strength. A further increment in the instrument rotational speed somewhat lessens the tractable shear quality. In addition, for a steady device rotational speed, expanding the dwell time builds the elastic shear quality. Nevertheless, expanding the abide time over 6 s did not altogether build the pliable shear quality [35]. A most extreme pliable shear quality of 5.14 kN was displayed by the weld at a device rotational speed of 800 rpm and a stay time of 8 s.

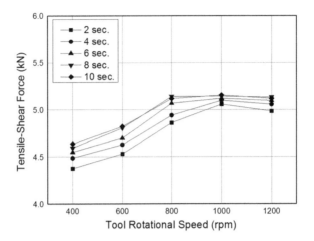

**FIGURE 8.10**
Variation of the ultimate force with the tool rotation speed at several dwell times [35]. (With kind permission from Springer Nature.)

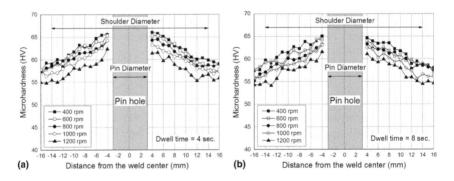

**FIGURE 8.11**
Vickers microhardness distributions of welds at different tool rotational speeds and constant dwell times of (a) 4 s and (b) 8 s [35]. (With kind permission from Springer Nature.)

Microhardness profiles are almost symmetric concerning the focal point of the pinhole. Generally, the stir zone showed higher small-scale hardness that esteems the base material. Smaller-scale hardness increments radially toward the pinhole. Expanding the tool rotational speed and additionally the dwell time somewhat reduces the hardness of the weld (see Figure 8.11) [35]. The lessening of the miniaturized scale hardness with the expanding tool rotational speed or potentially the stay time might be ascribed to the expansion of the grain estimate at the SZ in light of the Hall–Petch condition [37,38].

The compressive force associated with the axial movement of the tool during the plunging phase is referred to as the axial force or plunge rate [44,45]. An iron

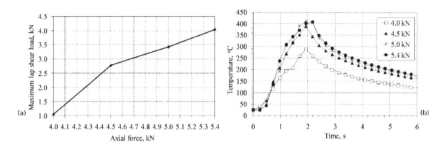

**FIGURE 8.12**
Lap-shear load and temperature results for varying axial force: (a) relationship between maximum lap-shear load to axial force and (b) temperature profile measured at specimen (2 mm away from shoulder) [46]. (With kind permission from Taylor & Francis.)

block underneath the lower sheet is utilized to help this axial force. Figure 8.12a demonstrates the most extreme lap-shear stack for different axial forces. For this situation, not at all like abide time and apparatus speed, the most extreme lap-shear stack expanded with expanding pivotal power. Figure 8.12b shows cases of average temperature profiles with varied axial forces.

The maximum temperature was the lowest for the lowest axial force (4.0 kN) but nearly identical for the other higher axial forces. The most extreme temperatures were come to at about a similar time for all cases. In a power controlled process, expanding the process duration brings about the shoulder entering further into the upper sheet and the joint line between the upper and lower sheet bends way up yonder, into the clouds from the free edge of the stick toward the free edge of the shoulder [46]. With higher axial forces, more vitality is moved into the example and the shoulder pierces into the upper sheet.

The device entrance and the dwell time basically decide the warmth age, material plasticization around the stick, weld geometry and thusly the mechanical properties of the weld joint [38]. The instrument descending power and the device rotational speed are kept up for a suitable time (dwell time) to create frictional warmth [38]. Bozkurt et al. [38] joined two different Al alloys sheets (AA 2024-T3 and 5754-H22) by FSSW employing different weld parameters including the apparatus tilt point. The impacts and effects of weld parameters such as plunge depth, dwell time, plate positioning and tool tilt angle on the mechanical properties were determined by keeping rotational speed and plunge rate constant. They utilized an apparatus turn speed of 1500 rpm with a steady dive rate of 3.5 mm/s to decide the impacts of the device abide time and the device dive profundity on the joint quality. Instrument dive profundities of 2.45, 2.55, and 2.65 mm from the example surface and stay times of 2, 5, and 10 s were utilized for the weld preliminaries. The apparatus tilt edge likewise differed from 0° to 3° (i.e., 1°, 2°, and 3°).

Lap-shear tensile properties (LS) of dissimilar joints produced using various welding conditions are presented in Figure 8.13. The lap-shear tensile properties increase with increasing tool plunge depth and dwell time for all conditions. As shown in Figure 8.13c, the maximum LS value, i.e., 5.36 kN, was obtained from the joints produced by placing AA 5754-H22 alloy as the upper sheet at zero tilt, a plunge depth of 2.65 mm and a dwell time of 10 s. Conversely, the lowest LS value, i.e., 1.79 kN, was obtained for the maximum tilt angle (3°), a tool plunge depth of 2.45 mm, and a dwell time of 2 s (Figure 8.13a). Higher dwell times lead to a larger weld zone, thus a higher joint strength, resulting from the intensive stirring action. However, the joint produced with zero tilt yielded the highest LS value when compared with the joint produced by placing the AA 5754 plate on top. This is possibly due to asymmetrical nugget formation at zero tilt. The motives for attaining diverse LS values for unalike joints are intricate. One of those motives is the different peak temperatures the material experiences during FSSW as a result of the various welding parameters. A differential is the temper condition of the material, whether it is age hardened or solution treated and strain hardened, which has a significant effect on the mechanical behavior of the joints. The plate positioning in dissimilar lap joints also plays a vital role owing to the fact that the tool shoulder acts on the top plate surface; thus, the degree of softening due to the heat input varies depending on the temper condition of the plate placed on the top. The age-hardened alloys in general experience a higher degree of softening in the weld zone during FSW or FSSW because of the dissolution and/or coarsening of the strengthening precipitates due to the annealing effect than those that have been solution treated and/or strain hardened. The optimum tool tilt angle was found to be 0° and 2° for AA 5754/2024 and AA 2024/5457 joints, respectively. A further increase in tool tilt angle decreased the strength of the joint due to an asymmetrical nugget formation resulting from insufficient stirring when tilting is employed.

## 8.5  Effect of Tool Design

The tooling pin and shoulder profiles are significant factors that influence the mechanical and metallurgical properties of the FSSW joints [47–51]. The effect of the tool design on the mechanical as well as metallurgical characteristics of friction stir spot welded joints was investigated by Badarinarayan et al. [49,50]. The tool geometries evaluated are presented in Figure 8.14. Both shoulder and pin profiles were varied.

The shoulder plunge depth was fixed at 0.2 mm. The shoulder plunge depth is circuitously measured by deducting the pin length from the total tool plunge depth. FSSW-C welds exhibit the highest strength, while FSSW-X welds have the lowest strength. The macroscopic cross section of one of the

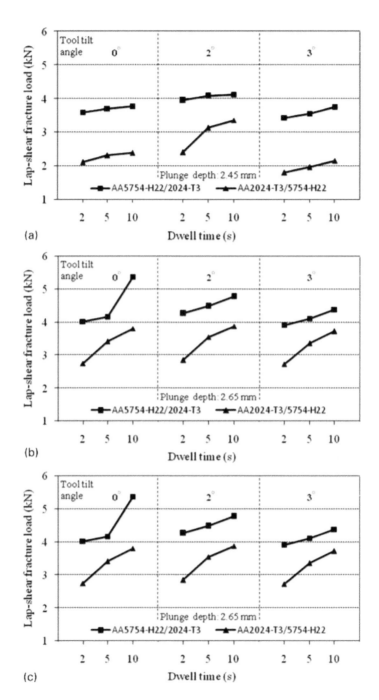

**FIGURE 8.13**
Effect of tool tilt angle and dwell time on LS of friction stir spot welded (a) 2.45, (b) 2.55, and (c) 2.65 mm [38]. (With kind permission from Taylor & Francis.)

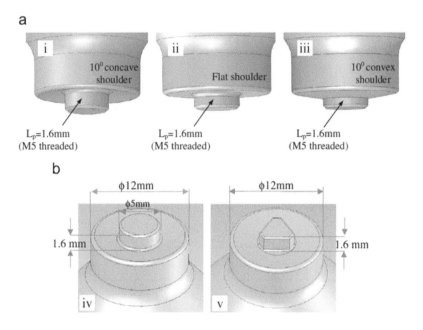

**FIGURE 8.14**

Schematic illustration of FSSW tool geometries: (a) cylindrical pin tools having (i) concave shoulder, (ii) flat shoulder, and (iii) convex shoulder. All pins have a length of 1.6mm and M5 threads (b) Concave shoulder profiles having (iv) cylindrical pin (1.6mm pin length and M5 threads) and (v) triangular pin (1.6mm pin length, no threads) [50]. (With kind permission from Elsevier.)

weld joints is presented in Figure 8.15. Several significant geometrical weld features are visible. These include effective thickness, hook height, hook width, and stir zone width.

Referring to Figure 8.16b,c, the ruptured sample pictures of the FSSW-C, as well as FSSW-F, disclose that the concluding fissure trail is through the operative thickness of the uppermost sheet. Consequently, $T_{eff}$ becomes significant in determining the welding strength as failure occurs through the top sheet. The effective thickness of the top sheet proffers opposition alongside exterior loading. As a result, in order to achieve welding with high static strength, one needs to have $T_{eff}$ as bulky as possible. FSSW-C demonstrated the largest $T_{eff}$ (therefore inferring the highest static strength), while FSSW-X holds the smallest $T_{eff}$ (therefore inferring the lowermost strength). The magnitude of $T_{eff}$ is a consequence of the intrinsic geometry of the shoulder. The concave shoulder offers an additional "pocket" of volume around the root of the pin [50]. This volume is first filled up by the extruded material before the shoulder comes in contact with the workpiece. Alternatively, the convex shoulder "sticks out" and is in touching base with the workpiece at a much prior plunge depth, if likened to the FSSW-C tool, thereby reducing the top sheet thickness sooner. The effect associated with the flat shoulder lies

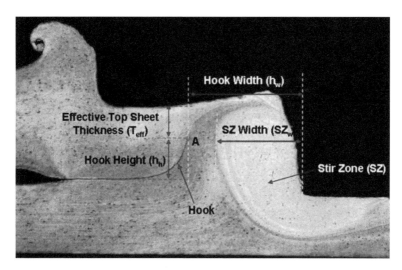

**FIGURE 8.15**
Cross-sectional macrostructure of a typical FSSW illustrating the different weld geometrical features [50]. (With kind permission from Elsevier.)

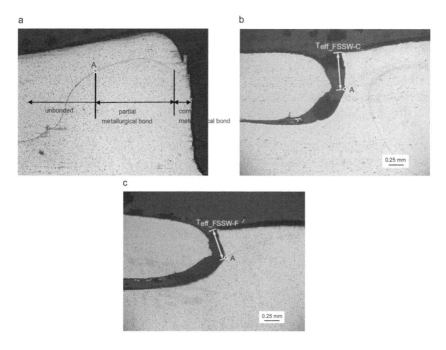

**FIGURE 8.16**
(a) Typical cross-sectional macrostructure of a friction stir spot weld showing different bonding regions: (b) fractured FSSW-C specimen (after external loading) and (c) fractured FSSW-F specimen (after external loading) [50]. (With kind permission from Elsevier.)

in between these two. The static strength differences between the diverse profiles of the shoulder may then be directly inferred from this behavior. The welds made with the FSSW-C tool revealed the occurrence of a constant hook and a larger stir zone in the weld area while matching the diverse tool pin profiles. The hook geometry originating at the crossing point of the two sheets persisted along the boundary of the stir zone and eventually ended very adjacent to the keyhole. On the contrary, welds carried out with the FSSW-T device exhibited a detained hook but a much smaller stir zone for similar plunge depths. The tool geometry, in particular, the pin profile, is a predominant aspect in defining the welding geometry and is, therefore, the static strength of these joints [49]. Cox et al. [51] showed that the tensile strength typically declines as the length of pin rises from no pin to a length commensurate with penetration to 10% of the bottom plate. Three distinctive letdown modes could be recognized when the welds were placed under tensile loading: mixed mode, shear mode, and finally the nugget-pullout mode.

## 8.6 Process Variants

### 8.6.1 Refilling Spot Welding

Friction spot welding (FSpW), also referred to as refill FSSW, happens to be a relatively new-fangled solid-state linking process that is a viable substitute for creating dissimilar overlap spot joints [52]. The procedure utilizes a non-consumable tool consisting of two movable parts—pin and sleeve—mounted coaxially to a clamping ring (see Figure 8.17a) [52]. A graphic diagram of the FSpW procedure is displayed in Figure 8.17b. In the initial stage, the higher and lower plates are secured by the fastening ring and the backing anvil before both the pin and the sleeve rotation commences. During the second stage, the pin and sleeve are displaced in opposite directions to each other; one is plunged into the material, while the other retracts from the surface

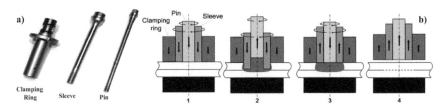

**FIGURE 8.17**
(a) Friction spot welding tool. (b) Illustration of the FSpW process using sleeve plunge variant: (1) clamping and tool rotation, (2) sleeve plunge and the pin retraction, (3) parts back to surface level, and (4) tool removal [56]. (With kind permission from Elsevier.)

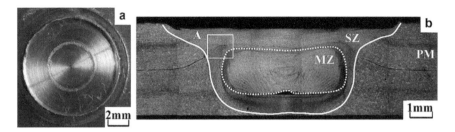

**FIGURE 8.18**
Macroscopic appearance of FSSW joint with refilled probe hole: (a) top view of the weld zone and (b) cross section of weld zone [52]. (With kind permission from Elsevier.)

to create a space (reservoir) where the overflow of plasticized material is housed. The rotating pin and sleeve generate sufficient frictional heating to plasticize a volume of material directly beneath the apparatus.

In the third stage, after getting at the pre-defined plunge profundity and dwell time, the pins, as well as a sleeve, are displaced toward the surface of the plate, thereby forcing the displaced material to refill the keyhole completely. In the last stage, the implement is detached from the surface of the plate, and a weld shorn of the keyhole is the result [53–55]. Figure 8.18 shows a top view and cross section of the weld zone in a typical joint with refilled probe hole. The successfully refilled probe hole at the instance of the double-acting tool is clearly visible [56].

## 8.6.2 Double-Side Spot Welding

An extra, revolving anvil analogous to a pin-less FSW implement may be utilized for FSSW of reedy metal plates. It is also useful to join denser cross sections, improve the mechanical force of the spot weld, and reduce the reaction forces on the spot welding mount. A typical double-sided FSSW device is presented in Figure 8.18. With the aim of maintaining the suitable positioning of the workpiece, and revolving anvil a "free-floating" stage is executed. The stage motion is steered by three cylindrical posts that rest on three die springs (see Figure 8.19). To offset the oscillatory attribute of the die springs, a huge stabilizing arm is fastened to the stage in an attempt to offer damping and stability. The opposite end of the stabilization arm is attached to a precision hinge to limit the motion of the sample stage to the Z-direction only. The contact between the welding tool and workpiece causes the die springs supporting the sample stage to be compacted. The compacting of the die springs facilitates the revolving anvil to come in contact with the workpiece. Once the spot weld is completed, the worktable is depressed, so as to return the die springs to their initial position, retracting the anvil from coming in contact with the workpiece [57].

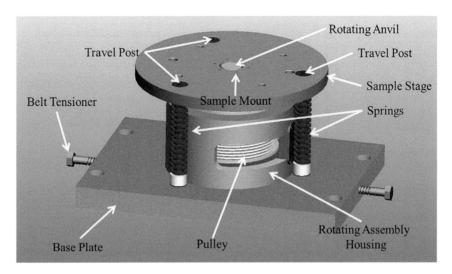

**FIGURE 8.19**
CAD drawing of the RAFSSW device. The workpieces are mounted to the sample stage. During welding, the die springs supporting the sample stage are compressed, allowing for both the welding tool and rotating anvil to be in contact with the weldment during welding [57]. (With kind permission from Elsevier.)

### 8.6.3 Pin-Less Tool

This is a recently developed technique that involves a process in which the plunging tool comprises a shoulder only that is applied to the upper plate. In this procedure, the chafing in the middle of the tool and the upper plate leads to plastic flow in the upper plate and speedy heating which, along with the pressure applied by the device, produces a joint (see Figure 8.20).

Notwithstanding the joining properties accomplished by spot rubbing mix welding, it has the accompanying advantages: (1) it doesn't make a gap relating to the probe, (2) no burr is made by the metal being removed from the opening, and (3) the welding time is short [58].

The pin-less tool gives good results because of a more uniform contact across the tool surface, thereby increasing the likelihood of efficient bonding beneath the shoulder at the verge of the weld. A pin-less device may also help to prevent hooking (see Figure 8.21), which may be triggered by the upward flow of the bottom sheet material displaced by the pin [59].

### 8.6.4 External Heating

Additional external heating may be applied to modify the FSSW process. Shen et al. [60] investigated the effect of heating the AZ31 magnesium alloy specimen during welding. Additional heat is applied by heating the backing

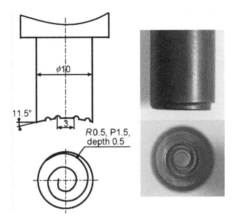

**FIGURE 8.20**
Configuration and a photograph of the newly developed tool. All dimensions are in millimeters [11]. (With kind permission from Elsevier.)

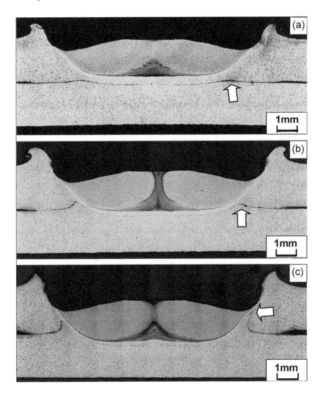

**FIGURE 8.21**
Weld structures of the longitudinal section for different shoulder plunge depths produced by scroll tool under processing condition of a tool rotation speed of 3000 rpm and a tool holding time of 4 s: 0.5, (b) 0.7, and (c) 0.9 mm. Arrows indicate the upward material flow of the lower sheet [11]. (With kind permission from Elsevier.)

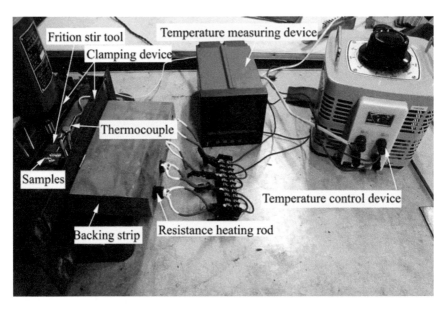

**FIGURE 8.22**
Schematic representation of the FSSW of AZ31 magnesium alloy [60].

strip with resistive elements. The backing strip temperature was varied. The results showed that external heating augmented the bond width and condensed the width of the partial metallurgical zone during FSSW welding of AZ31 magnesium alloy sheets. The additional heat increased the forming ability and fluidity of AZ31 magnesium alloy resulting in fewer spaces in the joints [60] (Figure 8.22).

### 8.6.5 Particle Addition

Graphite/aluminum metal matrix composite (MMC) joints may be efficaciously contrived by FSSW. In the course of lap-joining procedures of aluminum alloy sheets by FSSW, graphite/water colloid is applied sandwiched between the implement shoulder and the upper plate. A Raman spectrum of the FSSW joint endorses that the graphite/aluminum MMC is successfully introduced in the stir zone [61]. Quasi-static shear tests and microhardness measurements of these FSSW joints also show that the strength and toughness of the joint are significantly improved by introducing graphite/aluminum MMC in the stir zone. This implies that the mechanical properties of the structural joints by FSSW may be relatively easily enhanced by the introduction of MMC into the joints negating the need for possible material changes or increasing the number of joints [61].

## 8.7 Future Trends

### 8.7.1 Ferrous Alloys

Although FSSW was largely developed for use with nonferrous alloys, the technology may also be applied to ferrous alloys. Selected studies on FSSW of steel and other ferrous alloys are reported in [62–67]. Ahmed et al. investigated FSSW of high-Mn twinning-induced plasticity steel (TWIP). Macrographs depicting the top view of these joints for diverse rotational speeds are provided in Figure 8.23. The top outlook unmistakably shows perceived defect-free joints of high-Mn TWIP with a clear HAZ around the joint that rises in size with growing rotational speed. Figure 8.24 presents macrographs of the transverse cross section of the joints. It shows that the interface between the overlapped sheets is displaced upward due to the complex plasticized material flow produced by the rotation of the tool. Three discrete regions develop in the weld region once the gyrating shoulder comes in contact with the upper sheet. They are the flow transition zone beneath the tool shoulder, the stir zone around the probe, and the torsion zone beneath the tool probe. These three zones are clearly visible in the macrographs presented in Figure 8.24. Furthermore, it can be observed that the HAZ width increases with increasing rotation speed, as indicated by the superimposed dashed lines. To clearly identify the three distinct regions in the weld area, micrographs of the joint at various locations are displayed in Figure 8.25. Figure 8.25a,b displays the stir zone that formed around the implement probe, whereas Figure 8.25c,d displays the torsion zone that formed beneath the instrument probe. It can be observed that both zones are very thin, around 200 μm thick, and display a dynamically recrystallized grain structure. Also, it can be noted

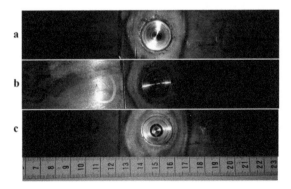

**FIGURE 8.23**
Macrographs of the top surface of the friction stir spot welded TWIP steel joints at (a) 500, (b) 750, and (c) 1000 rpm [62]. (With kind permission from Elsevier.)

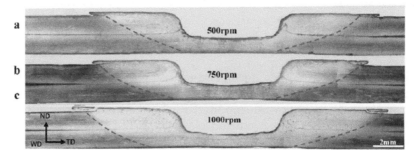

**FIGURE 8.24**
Optical macrographs of transverse cross section: (a) 500, (b) 750, and (c) 1000 rpm. Dashed lines show the HAZ width. ND is the normal direction, TD is the transverse direction, and WD is the welding direction of FSW axes [62]. (With kind permission from Elsevier.)

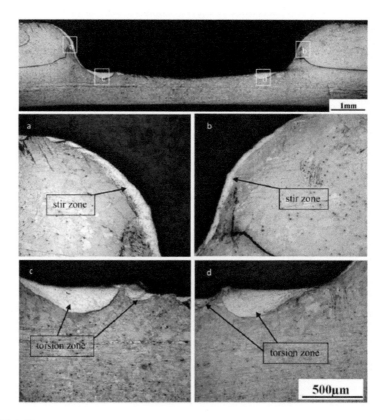

**FIGURE 8.25**
Optical microstructures of the FSSW of TWIP steel at a rotation speed of 750 rpm for different regions: (a) and (b) at the top adjacent to the surface of the tool probe; (c) and (d) underneath the tool probe [62]. (With kind permission from Elsevier.)

that the stir zone is almost continuous around the tool probe of the same thickness, while the torsion zone has an uneven thickness.

FSSWs of galvanized steel [63], low-carbon steel [63,67], mild steel [65], and austenitic stainless steel [67] were also all conducted successfully. However, the design and development of the tool for joining such ferrous alloys are challenging and costly. Because of the limited tool penetration and the reduced pin height of the welding tool (less than plate thickness), the extent of the mechanically mixed layer between the top and bottom plates at the weld nugget is nonexistent or limited [63]. FSSW of low-carbon steel requires the introduction of an additional heat source for preheating to achieve sound welds with acceptable loads and rotational speeds [66]. When these challenges are overcome, FSSW of ferrous metals may become a more mainstream and show greater promise.

## 8.7.2 Nonmetals

Modern thermoplastic constituents are applied in an ever-expanding assortment of engineering uses due to their improved strength-to-weight ratios, robustness, fast solidification, and low thermal conductivity. FSSW of nonmetals has been successfully conducted. It was found that welding parameters such as tool rotational speed, plunge depth, and dwell time play an important role in controlling weld strength [68–72]. The appearance of friction stir spot welded polypropylene sheeting is presented in Figure 8.26. Joining was achieved without visible deformation of the upper or lower sheet. Essentially, the tool shape is transferred onto the joint and the excess base material is displaced to form a protruding ring around the weld.

The macrostructures of a friction stir spot welded joint of polyethylene are displayed in Figure 8.27. The weld nugget thickness (x) is clearly visible. The weld nugget thickness is an indicator of the weld bond area and therefore also a direct indicator of the strength of the FSSW joint.

**FIGURE 8.26**
Appearance of a friction stir spot weld joint of polypropylene [69]. (With kind permission from Elsevier.)

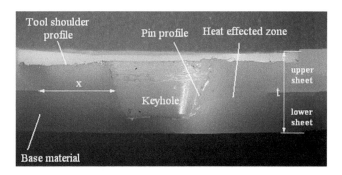

**FIGURE 8.27**
Macrostructure of the FSSW joint of polyethylene sheets. x: nugget thickness and t: material thickness [73]. (With kind permission from Elsevier.)

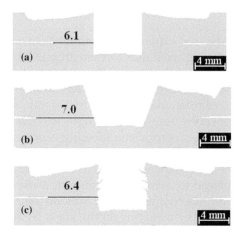

**FIGURE 8.28**
Effect of pin angle on weld nugget formation: (a) straight cylindrical pin, (b) 15° pin angled tapered cylindrical pin, and (c) threaded cylindrical pin [73]. (With kind permission from Elsevier.)

The size of the keyhole that forms in the weld zone (polyethylene) is a direct function of the pin profile (see Figure 8.28). Three different pin profiles were tested (a straight, an angled, and a threaded pin) [71]. The keyhole shape corresponds almost exactly with the pin profile (see Figure 8.28).

To obtain an optimum weld macrostructure for adequate strength, the process needs to be optimized for apparatus rotational speed, tool plunge depth, and dwell time along with an appropriate pin geometry. The tool pin geometry has a substantial effect on the nugget thickness and therefore the weld strength. Optimum welding parameters induce sufficient melting of the high-density polyethylene in the HAZ. Extraordinary value welds with optimum strength are then the result. The delay time, tool pin angle, shoulder concavity angle, and the shoulder diameter have a significant effect on the nugget thickness and therefore on the weld strength of the joint. The best weld strength was obtained at the highest welding force [71]. This study clearly illustrated that selected nonmetallic materials may also be successfully joined by FSSW.

## 8.8 Summary

The current chapter briefly introduced the relevant technology associated with FSSW. The discussions were largely based on current published data. The joining mechanism and the microstructural evolution including material flow, hook formation, local melting, cracking, and formation of intermetallic compounds in the joints were discussed. The perceived effects of process parameters including rotational speed, dwell time, plunge depth, plunge rate, and tool tilting angle on the mechanical and metallurgical properties were presented. The influences of tool geometry and tool profile on the joint properties were also presented. Latest developments to refill the exit hole and the technology of double-side spot welding were described. The use of pin-less tool designs, the technique of external heat addition, and the introduction of MMCs during FSSW were also introduced. The chapter was concluded with a brief discussion highlighting future trends associated with FSSW of ferrous and nonmetals.

## References

1. Tran, V. X., and Pan, J. 2010. Analytical stress intensity factor solutions for resistance and friction stir spot welds in lap-shear specimens of different materials and thicknesses. *Engineering Fracture Mechanics* 77:2611–2639.
2. Chao, Y. J., Qi, X., and Tang, W. 2003. Heat transfer in friction stir welding—experimental and numerical studies. *Journal of Manufacturing Science Engineering* 125:138–145.
3. ThoppulSrinivasa, D., and Gibso Ronald, F. 2009. Mechanical characterization of spot friction stir welded joints in aluminum alloys by combined experimental/numerical approaches. *Materials Characterisation* 60:1342–1351.
4. Wang, D. A., Chao, C. W., Lin, P. C., and Uan, J. Y. 2010. Mechanical characterization of friction stirs spot micro welds. *Journal of Materials Processing Technology* 210:1942–1948.
5. Khan, M. I., Kuntz, M. L., Su, P., Gerlich, A., North, T., and Zhou, Y. 2007. Resistance and friction stir spot welding of DP600: a comparative study. *Science and Technology of Welding and Joining* 12:175–182.

6. Yuan, W. 2008. Friction stir spot welding of aluminum alloys. *Masters Theses.* 5429. http://scholarsmine.mst.edu/masters_theses/5429.

7. Awang, M., and Mucino, V. H. 2010. Energy generation during friction stir spot welding (FSSW) of Al 6061-T6 Plates. *Materials and Manufacturing Processes* 25:167–174.

8. Fanelli, P., Vivio, F., and Vullo, V. 2012. Experimental and numerical characterization of Friction Stir Spot Welded joints. *Engineering Fracture Mechanics* 81:17–25.

9. Pathak, N., Bandyopadhyay, K., Sarangi, M., and Panda, S. K. 2013. Microstructure and mechanical performance of friction stir spot-welded aluminum-5754 sheets. *Journal of Materials Engineering and Performance* 22:131–144.

10. Fujimoto, M., Koga, S., Abe, N., Sato, S. Y., and Kokawa, H. 2009. Analysis of plastic flow of the Al alloy joint produced by friction stir spot welding. *Welding International* 23:589–596.

11. Tozaki, Y., Uematsu, Y., and Tokaji, K. 2010. A newly developed tool without probe for friction stir spot welding and its performance. *Journal of Materials Processing Technology* 210:844–851.

12. Godhani, P. S., Patel, V., Vora, J. J., Chaudhary, N. D., and Banka, R. 2019. Effect friction stir welding of aluminum alloys AA6061/AA7075: Temperature measurement, microstructure, and mechanical properties. *Innovations in Infrastructure* 591–598 doi: 10.1007/978-981-13-1966-2_53.

13. Shen, Z., Yang, X., Zhang, Z., Cui, L., and Yin, Y. 2013. Mechanical properties and failure mechanisms of friction stir spot welds of AA 6061-T4 sheets. *Materials and Design* 49:181–191.

14. Yang, Q., Mironov, S., Sato, Y. S., and Okamoto, K. 2010. Material flow during friction stir spot welding. *Materials Science and Engineering* 527:4389–4398.

15. Su, P., Gerlich, A., North, T. H., and Bendzsak, G. J. 2007. Intermixing in dissimilar friction stir spot welds. *Metallurgical and Materials Transactions* 38:584–595.

16. Yuan, W., Mishra, R. S., Carlson, B., Verma, R., and Mishra, R. K. 2012. Material flow and microstructural evolution during friction stir spot welding of AZ31 magnesium alloy. *Materials Science and Engineering* 543:200–209.

17. Su, P., Gerlich, A., North, T. H., and Bendzsak, G. J. 2006. Material flow during friction stir spot welding. *Science and Technology of Welding and Joining* 11:61–71.

18. Song, X., Ke, L., Xing, L., Liu, F., and Huang, C. 2014. Effect of plunge speeds on hook geometries and mechanical properties in friction stir spot welding of A6061-T6 sheets. *International Journal of Advanced Manufacturing Technology* 71:2003–2010.

19. Yin, Y. H., Sun, N., North, T. H., and Hu, S. S. 2010. Hook formation and mechanical properties in AZ31 friction stir spot welds. *Journal of Materials Processing Technology* 210:2062–2070.

20. Su, P., Gerlich, A., Yamamoto, M., and North, T. H. 2007. Formation and retention of local melted films in AZ91 friction stir spot welds. *Journal of Materials Science* 42:9954–9965.

21. Yang, Y. K., Dong, H., Cao, H., Chang, Y. A., and Kou, S. 2008. Liquation of Mg alloys in friction stir spot welding. *Welding Journal* 87:168–177.

22. Gerlich, A., Yamamoto, M., and North, T. H. 2007. Local melting and cracking in Al 7075-T6 and Al 2024-T3 friction stir spot welds. *Science and Technology of Welding and Joining* 12:472–480.

23. Gerlich, A., Yamamoto, M., and North, T. H. 2008. Local melting and tool slippage during friction stir spot welding of Al-alloys. *Journal of Materials Science* 43:2–11.
24. Yamamoto, M., Gerlich, A., North, T. H., and Shinozaki, K. 2008. Cracking and local melting in Mg-alloy and Al-alloy during friction stir spot welding. *Welding in the World* 52:9–10.
25. Yamamoto, M., Gerlich, A., North, T. H., and Shinozaki, K. 2008. Cracking in dissimilar Mg alloy friction stir spot welds. *Science and Technology of Welding and Joining* 13:583–592.
26. Yamamoto, M., Gerlich, A., North, T. H., and Shinozaki, K. 2007. Cracking in the stir zones of Mg-alloy friction stir spot welds. *Journal of Materials Science* 42:7657–7666.
27. Rao, H. M., Yuan, W., and Badarinarayan, H. 2015. Effect of process parameters on mechanical properties of friction stir spot welded magnesium to aluminium alloys. *Materials and Design* 66:235–245.
28. Shen, J., Suhuddin, U. F. H., Cardillo, M. E. B., and Santos, J. F. D. 2014. Eutectic structures in friction spot welding joint of aluminum alloy to copper. *Applied Physics Letters* 104:1919011–1919014.
29. Mubiayi, M. P., and Akinlabi, E. T. 2016. Evolving properties of friction stir spot welds between AA1060 and commercially pure copper C11000. *Transaction of Nonferrous Metal Society of China* 26:1852–1862.
30. Choi, D. H., Ahn, B. W., Lee, C. Y., Yeon, Y. M., Song, K., and Jung, S. B. 2011. Formation of intermetallic compounds in Al and Mg alloy interface during friction stir spot welding. *Intermetallics* 19:125–130.
31. Bozzi, S., Etter, A. L. H., Baudin, T., Criqui, B., and Kerbiguet, J. G. 2010. Intermetallic compounds in Al 6016 IF-steel friction stir spot welds. *Materials Science and Engineering* 527:4505–4509.
32. Karthikeyan, R., and Balasubramanian, V. 2010. Predictions of the optimized friction stir spot welding process parameters for joining AA2024 aluminum alloy using RSM. *International Journal of Advanced Manufacturing Technology* 51:173–183.
33. Merzoug, M., Mazari, M., Berrahal, L., and Imad, A. 2010. Parametric studies of the process of friction spot stir welding of aluminium 6060-T5 alloys. *Materials and Design* 31:3023–3028.
34. Shiraly, M., Shamanian, M., Toroghinejad, M. R., and Ahmadi Jazani, M. 2014. Effect of tool rotation rate on microstructure and mechanical behavior of friction stir spot-welded AlCu composite. *Journal of Materials Engineering Performance* 23:413–420.
35. Mahmoud, T. S., and Khalifa, T. A. 2014. Microstructural and mechanical characteristics of aluminum alloy AA5754 friction stir spot welds. *Journal of Materials Engineering Performance* 23:898–905.
36. Gerlich, A., Cingara, G. A., and North, T. H. 2006. Stir zone microstructure and strain rate during Al 7075-T6 friction stir spot welding. *Metallurgical and Materials Transactions* 37:2773–2785.
37. Barlas, Z. 2015. Effect of friction stir spot weld parameters on Cu CuZn30 bimetal joints. *International Journal of Advanced Manufacturing Technology* 80:161–170.
38. Bozkurt, Y., Salman, S., and Cam, G. 2013. Effect of welding parameters on lap shear tensile properties of dissimilar friction stir spot welded AA 5754-H22 2024-T3 joints. *Science and Technology of Welding and Joining* 18:337–345.

39. Zhang, Z., Yang, X., Zhang, J., Zhou, G., Xu, X., and Zou, B. 2011. Effect of welding parameters on microstructure and mechanical properties of friction stir spot welded 5052 aluminum alloy. *Materials and Design* 32:4461–4470.

40. Gerlich, A., Su, P., Yamamoto, M., and North, T. H. 2007. Effect of welding parameters on the strain rate and microstructure of friction stir spot welded 2024 aluminum alloy. *Journal of Materials Science* 42:5589–5601.

41. Lin, Y. C., Liu, J. J., Lin, B. Y., Lin, C. M., and Tsai, H. L. 2012. Effects of process parameters on strength of Mg alloy AZ61 friction stir spot welds. *Materials and Design* 35:350–357.

42. Fujimoto, M., Watanabe, D., Abe, N., Yutaka, S. S, and Kokawa, H. 2010. Effects of process time and thread on tensile shear strength of Al alloy lap joint produced by friction stir spot welding. *Welding International* 24:169–175.

43. Tran, V. X., Pan, J., and Pan, T. 2009. Effects of processing time on strengths and failure modes of dissimilar spot friction welds between aluminium 5754-O and 7075-T6 sheets. *Journal of Materials Processing Technology* 209:3724–3739.

44. Jonckheere, C., Meester, B. D., Cassiers, C., Delhaye, M., and Simar, A. 2012. Fracture and mechanical properties of friction stir spot welds in 6063-T6 aluminum alloy. *International Journal of Advanced Manufacturing Technology* 62:569–575.

45. Lathabai, S., Painter, M. J., Cantin, G. M. D., and Tyagi, V. K. 2006. Friction spot joining of an extruded Al–Mg–Si alloy. *Scripta Materialia* 55:899–902.

46. Arul, S. G., Miller, S. F., Kruger, G. H., Pan, T. Y., Mallick, P. K., and Shih, A. J. 2008. Experimental study of joint performance in spot friction welding of 6111-T4 aluminium alloy. *Science and Technology of Welding and Joining* 13:629–637.

47. Yuan, W., Mishra, R. S., Webb, S., Chen, Y. L., Carlson, B., Herling, D. R., and Grant, G. J. 2011. Effect of tool design and process parameters on properties of Al alloy 6016 friction stir spot welds. *Journal of Materials Processing Technology* 211:972–977.

48. Tozaki, Y., Uematsu, Y., and Tokaji, K. 2007. Effect of tool geometry on microstructure and static strength in friction stir spot welded aluminium alloys. *International Journal of Machine Tools & Manufacture* 47:2230–2236.

49. Badarinarayan, H., Yang, Q., and Zhu, S. 2009. Effect of tool geometry on static strength of friction stir spot-welded aluminum alloy. *International Journal of Machine Tools & Manufacture* 49:142–148.

50. Badarinarayan, H., Shi, Y., Li, X., and Okamoto, K. 2009. Effect of tool geometry on hook formation and static strength of friction stir spot welded aluminium 5754-O sheets. *International Journal of Machine Tools & Manufacture* 49:814–823.

51. Cox, C. D., Gibson, B. T., Strauss, A. M., and Cook, G. E. 2012. Effect of pin length and rotation rate on the tensile strength of a friction stir spot-welded Al alloy: a contribution to automated production. *Materials and Manufacturing Processes* 27:472–478.

52. Uematsu, Y., Tokaji, K., Tozaki, Y., Kurita, Y., and Murata, S. 2008. Effect of refilling probe hole on tensile failure and fatigue behaviour of friction stir spot welded joints in Al–Mg–Si alloy. *International Journal of Fatigue* 30:1956–1966.

53. Zhang, Z., Wang, X., Wang, P., and Zhao, G. 2014. Friction stir keyhole less spot welding of AZ31 Mg alloy-mild steel. *Transaction of Nonferrous Metal Society of China* 24:1709–1716.

54. Rosendo, T., Parra, B., Tier, M. A. D., Da Silva, A. A. M., Dos Santos, J. F., Strohaecker, T. R., and Alcântara, N. G. 2011. Mechanical and micro structural

investigation of friction spot welded AA6181-T4 aluminium alloy. *Materials and Design* 32:1094–1100.

55. Shen, Z., Yang, X., Zhang, Z., Cui, L., and Li, T. 2013. Microstructure and failure mechanisms of refill friction stir spot welded 7075-T6 aluminum alloy joints. *Materials and Design* 44:476–486.

56. Plaine, A. H., Gonzalez, A. R., Suhuddin, U. F. H., Dos Santos, J. F., and Alcantara, N. G. 2015. The optimization of friction spot welding process parameters in AA6181-T4 and Ti6Al4V dissimilar joints. *Materials & Design* 83:36–41.

57. Cox, C. D., Gibson, B. T., DeLapp, D. R., Strauss, A. M., and Cook, G. E. 2014. A method for double-sided friction stir spot welding. *Journal of Manufacturing Processes* 16:241–247.

58. Aota, K., and Ikeuchi, K. 2009. Development of friction stir spot welding using rotating tool without probe and its application to low-carbon steel plates. *Welding International* 23:572–580.

59. Bakavos, D., Chen, Y., Babout, L., and Prangnell, P. 2011. Material interactions in a novel pinless tool approach to friction stir spot welding thin aluminum sheet. *Metallurgical and Materials Transactions* 42:1266–1282.

60. Shen, J., Min, D., and Wang, D. 2011. Effects of heating process on the microstructures and tensile properties of friction stir spot welded AZ31 magnesium alloy plates. *Materials and Design* 32:5033–5037.

61. Jeon, C., Jeong, Y., Hong, S., Hasan, M. D. T., Tien, H. N., Hur, S., and Kwon, Y. 2014. Mechanical properties of graphite aluminum metal matrix composite joints by friction stir spot welding. *Journal of Mechanical Science and Technology* 28:499–504.

62. Ahmed, M. M. Z., Essam Ahmed, E., Hamada, A. S., Khodir, S. A., El-Sayed Seleman, M. M., and Wynne, B. P. 2016. Microstructure and mechanical properties evolution of friction stir spot welded high-Mn twinning-induced plasticity steel. *Materials and Design* 91:378–387.

63. Baek, S. W., Choi, D. H., Lee, C. Y., Ahn, B. W., Yeon, Y. M., and Song, K. 2015. Microstructure and mechanical properties of friction stir spot welded galvanized steel. *Materials Transactions* 51:1044–1050.

64. Baek, S. W., Choi, D. H., Lee, C. Y., Ahn, B. W., Yeon, Y. M., and Song, K. 2010. Structure–properties relations in friction stir spot welded low carbon steel sheets for light weight automobile body. *Materials Transactions* 51:399–403.

65. Sun, Y. F., Fujii, H., Takaki, N., and Okitsu, Y. 2012. Microstructure and mechanical properties of mild steel joints prepared by a flat friction stir spot welding technique. *Materials and Design* 37:384–392.

66. Sun, Y. F., Shen, J. M., Morisada, Y., and Fujii, H. 2014. Spot friction stir welding of low carbon steel plates preheated by high frequency induction. *Materials and Design* 54:450–457.

67. Jeon, J., Mironov, S., Sato, Y. S., Kokawa, H., Park, S. H. C., and Hirano, S. 2011. Friction stir spot welding of single-crystal austenitic stainless steel. *Acta Materialia* 59:7439–7449.

68. Lakshminarayanan, A. K., Annamalai, V. E., and Elangovan, K. 2015. Identification of optimum friction stir spot welding process parameters controlling the properties of low carbon automotive steel joints. *Journal of Materials Research Technology* 4:262–272.

69. Bilici, M. K. 2012. Application of Taguchi approach to optimize friction stir spot welding parameters of polypropylene. *Materials and Design* 35:113–119.

70. Patel, V., Sejani, D. J., Patel, N. J., Vora, J. J., Gadhvi, B. J., Padodara, N. R., and Vamja, C. D. 2016. Effect of tool rotation speed on friction stir spot welded AA5052-H32 and AA6082-T6 dissimilar aluminum alloys. *Metallography, Microstructure, and Analysis* 5 (2):142–148.
71. Goushegir, S. M., Dos Santos, J. F., and Amancio-Filho, S. T. 2015. Influence of process parameters on mechanical performance and bonding area of AA2024 carbon fiber reinforced poly friction spot single lap joints. *Materials & Design* 83:431–442.
72. Oliveira, P. H. F., Amancio-Filho, S. T., Santos, J. F. D., and Hage Jr, E. 2010. Preliminary study on the feasibility of friction spot welding in PMMA. *Materials Letters* 64:2098–2101.
73. Bilici, M. K., and Yukler, A. I. 2012. Influence of tool geometry and process parameters on macrostructure and static strength in friction stir spot welded polyethylene sheets. *Materials and Design* 33:145–152.

# 9

## *Linear Friction Welding*

**Xinyu Wang, Wenya Li, and Tiejun Ma**
*Northwestern Polytechnical University*

**Achilles Vairis**
*Northwestern Polytechnical University*
*TEI of Crete*

### CONTENTS

9.1 Linear Friction Welding Process ................................................................ 191
9.2 Advantages and Disadvantages of LFW ............................................... 192
9.3 LFW of Titanium Alloys ......................................................................... 193
    9.3.1 Macroscopic Features of Joints ................................................... 193
    9.3.2 Weld Microstructure of Joints .................................................... 194
    9.3.3 Texture Formation in Welds ........................................................ 197
    9.3.4 Residual Stresses of Joints .......................................................... 199
    9.3.5 Hardness Profiles of Joints .......................................................... 201
    9.3.6 Tensile Properties of Joints ......................................................... 205
    9.3.7 Impact Toughness, Fracture Toughness, Fatigue
           Properties, and Torsional Strength of Joints .............................. 205
9.4 Summary .................................................................................................... 206
References ............................................................................................................ 207

## 9.1 Linear Friction Welding Process

Linear friction welding (LFW) is a relatively new solid-state welding process where frictional heat is produced on the interface by a relative linear reciprocating movement between two components under a suitable combination of pressure, amplitude, and frequency of oscillation, as shown in Figure 9.1a. LFW is a self-regulating process, which means that it undergoes various phases of the process, wherein certain stress conditions must be reached at the welding interface and the nearby region. This process consists of four distinct phases, namely, initial phase, transition phase, equilibrium phase, and deceleration (or forging) phase, as schematically illustrated in Figure 9.1b.[1–3] Characterization of this process is based on observation with Ti-6Al-4V, and it is expected that all metals and their alloys follow it.

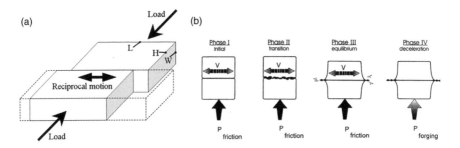

**FIGURE 9.1**
Schematic of the LFW process (a) and its distinct phases (b).[2] (With kind permission from Elsevier.)

Phase I – initial phase: Two components are brought into contact under pressure. The components' surfaces are rough, and heat is generated on the interface by solid friction. The true contact area increases with the softening and deformation of interface materials significantly. During this stage, no axial shortening of the specimens can be observed.

Phase II – transition phase: Large wear particles begin to extrude from the interface after interface materials are soften by sufficient frictional heat produced in Phase I. The true contact area increases to 100% of the cross-sectional area, and the soft plasticized layer formed on the interface is no longer able to support the axial load.

Phase III – equilibrium phase: Axial shortening of the components occurs at a nearly constant rate, accompanied by the formation of flash, due to the extrusion of plastic materials from the interface caused by the oscillatory movement.

Phase IV – deceleration phase: Once the desired upset is reached, the oscillatory movement is stopped very rapidly (within less than 0.1 s), and forging pressure may be applied on two components to form a sound joint.[2,3]

## 9.2 Advantages and Disadvantages of LFW

Unlike conventional fusion welding methods, LFW has a lot of distinctive advantages. The process is a solid-state one, and therefore solidification problems (e.g., hot cracking, porosity, segregation) can be avoided.[4] In the majority of cases, because of the intense deformation of the weld region, the weld presents a refined microstructure, with excellent mechanical properties relative to the parent material (PM). During LFW, the faying surfaces are kept tightly in contact; hence, no shielding gas or filler wire is required. In addition, the process has a self-cleaning ability owing to wear, with the wear debris being removed during the early stage of the process, and the plastic material is moved to flash during oscillatory motion. As the process does not use rotation, it is not limited to nonaxisymmetric components. The other advantages of LFW are the inherent stability and repeatability of the whole process, a high welding speed with the total weld time being less than 10 s for Ti64,[5,6] and similar high production rates (for AISI 316L and carbon manganese steel, the welding cycle time from start to finish is about 2 min). The iron in the speed advantage of the process is that it can become the inverse bottleneck in a factory as a single LFW machine needs a number of production lines to service in an efficient manner. Finally, high welding integrity can be achieved by LFW, in which the welding area can reach approximately 100% of the cross-sectional area of the joined parts.

On the other hand, the process shows some disadvantages, with the most important of these being the high welding equipment cost, limiting its industrial applications. LFW is being used commercially in aircraft engine manufacturer, in part owing to this. Another important limitation is that LFW machines need to be strongly constructed to contain parasitic vibrations due to the oscillatory movement of large metal masses. Thus, large and expensive fixed installations are necessary, and the development of cheap and portable machines is restricted by these design limitations.

## 9.3 LFW of Titanium Alloys

LFW is an effective welding technique for joining titanium alloys due to their good high-temperature properties like compressive yield and shear strength, and low thermal conductivities.[7] This allows the generated heat to remain at the interface causing the interface to rapidly heat and plasticize. As a result, more than half investigations focus on the LFW titanium alloys joints.[8]

### 9.3.1 Macroscopic Features of Joints

During the equilibrium phase of the process, provided the frictional heat input is sufficient, plastic material from across the whole rubbing interface is extruded to form a flash, which is especially pronounced for titanium alloys. This is a type of LFW, and it is usually associated with the quality of the weld. In all published studies on titanium alloys, the macroscopic features of the welds are similar. Successful titanium welds show appreciable flash from all sides of the joint with the flash length being larger in the direction of movement. The periodical ridges are attributed to the reciprocating motion and the extrusion of plasticized material during parts of the oscillatory cycle.

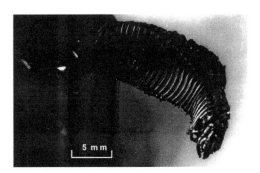

**FIGURE 9.2**
Detail of ridged flash of Ti64 linear friction weld.[2] (With kind permission from Elsevier.)

Vairis and Frost[2] studied macroscopically the Ti64 flash, where the plastic material yield dependence on temperature was verified through macroscopic investigation of the flash (Figure 9.2). The flash had a nonuniform thickness and appeared as ridges. Extruded matter from the sides of the specimen parallel to the direction of movement also showed the same characteristics. Due to no movement of the specimens in that direction, it may be assumed that axial shortening proceeds in a step-by-step mode during LFW. This pumping action causes yield and extrusion of material in pulses and confines the heat-affected zone (HAZ) close to the interface continuously.

The flash characteristics of other titanium alloys, such as Ti-6.5Al-1.5Zr-3.5Mo-0.3Si (Ti11),[9] Ti-5Al-2Sn-2Zr-4Mo-4Cr (Ti17),[10] and Ti-5Al-5V-5Mo-3Cr (Ti-5553),[11] have also been studied (Figure 9.3). From visual observation of the joints, the flash is the similar to that of Ti64[2] with material extruding from all four sides of the joint.

### 9.3.2 Weld Microstructure of Joints

The weld of Ti64 can be divided into three regions[5,12] according to their microstructure: the PM, the thermo-mechanically affected zone (TMAZ),

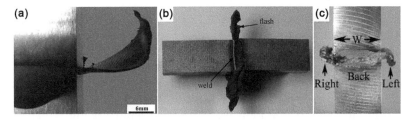

**FIGURE 9.3**
Photos of joints of (a) Ti11,[9] (b) Ti17[10], and (c) Ti5553.[11] (With kind permission from Springer, John Wiley and Sons and Elsevier.)

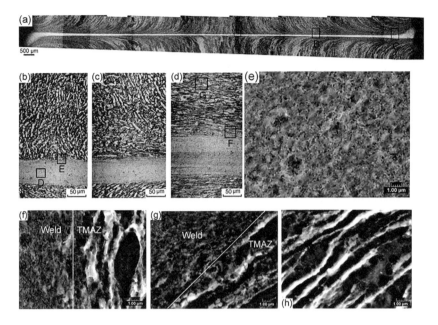

**FIGURE 9.4**
Cross-sectional microstructure of an LFW Ti64 sample along the oscillation direction. (a) Overall view of weld (OM); (b, c, and d) A, B, and C zones, respectively, marked in (a) (OM); (e and f) D and E zones, respectively, in (b) [Scanning electron microscope (SEM)]; (g and h) F and G zones, respectively, in (d) (SEM).[6] (With kind permission from John Wiley and Sons.)

and the weld center zone (WCZ), as shown in Figure 9.4. The characteristics of each zone are described as follows.

The Ti64 PM consists of a bimodal $\alpha+\beta$ structure in the form of alternating layers of equiaxed/globular $\alpha$ grains with transformed $\beta$ grains having a lamellar $\alpha+\beta$ appearance and elongated $\alpha$ grains with intergranular $\beta$, owing to the mill-annealed condition of the as-received material.[5]

For the linear friction-welded Ti64, the TMAZ consists of a severely deformed $\alpha+\beta$ microstructure owing to plastic deformation, which is oriented along the deformation direction.[13] Sun et al.[14] found the existence of elongated $\alpha$ with intergranular $\beta$ in the direct-drive friction-welded Ti64 welds, and a large number of dislocations in the $\alpha$ phase can also be found in the transmission electron microscope (TEM) photograph. There were no phase transformations observed in TMAZ, indicating that the temperature reached in this region does not exceed the $\beta$-transus temperature.

The WCZ is close to the weld line and the structure in this region is considerably different from that in PM; a Widmanstätten structure can be recognized, and its boundaries are delineated by prior-$\beta$ grains. Meanwhile, the initial bimodal $\alpha+\beta$ microstructure is disappeared. For the $\alpha+\beta$ Ti64 alloy, this change in microstructure can be explained by the fact that the $\beta$-transus temperature of 995°C has been exceeded. The presence of a Widmanstätten structure indicates that a complete $\alpha \to \beta$ transformation has occurred in the highest temperature region, i.e., WCZ.[5,15]

Sometimes, a martensitic structure also can be found in WCZ.[16] The difference in microstructure is due to the cooling rate from the single $\beta$-phase. If the weld cools at a rate faster than 410°C/s, then a diffusionless transformation occurs resulting in martensite. Some metastable $\beta$-phase may also remain. If the weld cools at a rate slower than 410°C/s, then a diffusional transformation occurs resulting in a Widmanstätten morphology.[17]

Ma et al.[13] found that the HAZ was limited whose borders cannot be determined in detail because of the stability of Ti64 at temperatures below 800°C. However, a HAZ is more noticeable in welds that are produced with low rubbing velocities.[18]

Compared to Ti64, the microstructural regions of other titanium alloy joints, such as CP Ti,[19] Ti11,[9] Ti17,[10,20] Ti6246,[21] and Ti-5553,[11] are very similar, although they have different $\beta$-transus temperatures. There are some common microstructural characteristics in the joints welded by various titanium alloys. Generally, the WCZ consists of fine recrystallized grains, and TMAZ is characterized by a mixed microstructure of deformed grains and partly recrystallized grains. However, due to the different alloy element contents in various titanium alloys, the characteristics of phases distribution in WCZ are various. For the $\alpha+\beta$ titanium alloys (such as Ti64, Ti11, Ti17, and Ti6246), martensite or $\alpha$ lath will be found in the recrystallized grains. However, for the single-phase titanium alloys (CP-Ti or Ti5553), only recrystallized grains can be observed without any precipitated phase.

It is clear from the papers published to date that there is no widely unified nomenclature for the microstructural regions in LFW, but HAZ, TMAZ, and WCZ are commonly used. Sometimes the plastically affected zone (PAZ) was also used to describe the region from one TMAZ/HAZ boundary to the other.[12,22,23]

In most cases, if the welding parameters are appropriate, the WCZ will presents a microstructure with fine recrystallized grains and good joining. However, in some cases, if the welding conditions cannot ensure a good extrusion of interface materials, some defects will present at the interface, such as oxide inclusions[5] and root flaw. More recently, Li et al.[24] found an interesting microstructure evolution phenomenon in LFW Ti64, i.e., abnormal spherical $\alpha$ phase presenting at the interface as shown in Figure 9.5, even with an axial shortening where the bonding should be good. Macroscopically, the bonding appeared like a crack, sometimes taking up more than 60% of the whole bondline (Figure 9.5a). When observing at high magnification (Figure 9.5b), the spherical $\alpha$ grains and the surrounding $\alpha'$ martensite located on $\beta$ matrix were clearly observed and there were almost no pores. These spherical $\alpha$ grains significantly decreased the joint tensile strength, just 44% of the PM and all the tensile fracture occurred at the bondline. Even after post-weld heat treatment (PWHT), these $\alpha$ grains could not be eliminated (Figure 9.5c). The tensile strength of the post-weld heat treated joint was just 71% of the PM

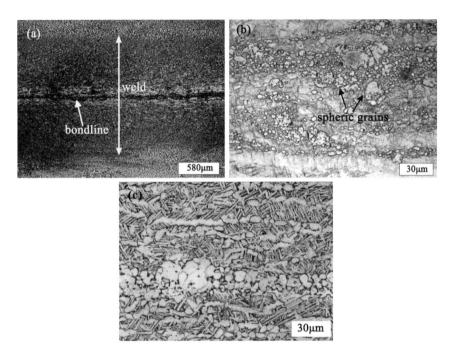

**FIGURE 9.5**
Cross-sectional microstructure of a LFW Ti64 joint along the oscillation direction. (a) OM image across the weld, (b) OM image of the bondline at high magnification, and (c) SEM image of weld center zone after PWHT.[24] (With kind permission from John Wiley and Sons.)

with the fracture occurring similarly at the bondline. They attributed this phenomenon to the microstructure of PM and the lower forging pressure. Recently, they also observed this phenomenon for LFW Ti11 titanium alloy.[25] The formation mechanism of the spherical $\alpha$ grains is similar to hot working, during which some lamellar $\alpha$ phase is broken up to form the spheroidal grains by boundary splitting. Subsequently, the recrystallization of the segments occurs and $\alpha$ lamellae transform into equiaxed grains. Under the applied friction pressure, the softer intergranular $\beta$ phase is extruded and small $\alpha$ particles amalgamate to form bigger ones.

### 9.3.3 Texture Formation in Welds

According to the published works, a texture is formed in the linear friction welded titanium alloy joint because of the heavy deformation occurring locally in the weld zone during welding as was studied with experiments. As expected texture generation in titanium alloys will significantly affect mechanical properties.

Karadge et al.[12] investigated the texture distribution in as-welded Ti64 joint and post-weld heat treated joint for the lab (small) and full (large)

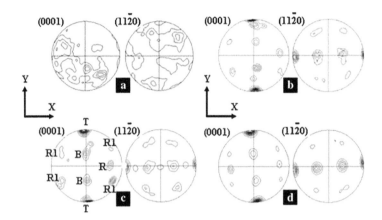

**FIGURE 9.6**
{10$\bar{1}$0}<11$\bar{2}$0> α textures in lab scale specimens: (a) base alloy (4X random), (b) as-welded (T: 35X, R: 2X, B: 5X, R1: 2X), (c) 1 h PWHT (T: 36X, R: 3X, B: 7X, R1: 5X), and (d) 8 h PWHT (T: 37X, R: 3X, B: 2X, R1: 1X) (it should be noted that XY is the weld plane.).[12] (With kind permission from Elsevier.)

scale specimens. They found an extremely strong transverse texture {10$\bar{1}$0}<11$\bar{2}$0> at WCZ in both sized specimens (Figure 9.6).[12] Furthermore, R-texture ({11$\bar{2}$2}<11$\bar{2}$3> was observed at a small distance away from the weld line in the large specimens. Both sized specimens showed weak basal- and R1-type textures. In addition, the effect of PWHT on the texture distribution was negligible.

Romero et al.[26] investigated the influence of welding pressure on texture development in LFW Ti64 joints. A strong transverse α texture {10$\bar{1}$0}<11$\bar{2}$0> could be observed at WCZ in the joint which was welded using a low applied pressure. With the increase of welding pressure, the texture intensity was reduced gradually.

More recently, Guo et al.[16] also investigated the texture variation in linear friction welded Ti64 joint. They found the texture changes gradually from WCZ to HAZ and pointed out that the number of transformed β had a significant effect on the resultant texture, but the effect of primary α grains on the texture formation in the weld was not notable.

For the β titanium alloy with BCC crystal structure, Dalgaard et al.[11] studied the LFW near-β titanium alloy Ti-5553 (Figure 9.7a). The results showed that WCZ consisted of fine recrystallized grains oriented with their <111> directions parallel to the welding oscillation direction.

As for the welding of pure Ti, Wang et al.[19] found that only the α phase with an evident {10$\bar{1}$0}<11$\bar{2}$0> texture existed in WCZ (Figure 9.7b). The presence of elongated grains in WCZ indicated incomplete recrystallization took place during LFW, and the texture variation did not change after PWHT.

The texture in LFW joints can be attributed to the combined effect of deformation and phase transformation during the welding process. When the

**FIGURE 9.7**
Inverse pole Figure maps of LFW Ti5553[11] (a) and pure Ti[19] (b). (With kind permission from Elsevier.)

materials are heated over the $\beta$-transus temperature, all $\alpha$ phase transforms to the $\beta$ phase and the plastic deformation occurs in the $\beta$ field. It has been widely accepted that the deformation mode of LFW can be simplified as simple shear. Thus, a {hkl}<111> or {112}<111> $\beta$ texture will be formed according to Gey and Humbert.[27] After cooling, the $\alpha$ phase will be precipitated from $\beta$ matrix keeping the Burgers orientation relationship and accompanying with the variant selection, and one or more $\alpha$ texture forms in WCZ in the end. However, the relationship between variant selection and welding parameters has not been clarified and more work in this field is necessary.

### 9.3.4 Residual Stresses of Joints

A number of different titanium alloys have been studied to measure residual stresses in completed joints. Residual stresses in LFW joints are caused by the thermal mismatch which is created by different cooling rates at various positions across the weld (e.g., material in the center of the weld will cool more slowly than at the surface) and the obvious microstructure change. The combination of thermomechanical deformation and the constraints to thermal expansion is the source of strain development during welding process.[26] The most frequently used methods for measurements of residual stresses[7,9,13,26,28,29] are high-energy synchrotron X-ray and neutron diffraction, and these methods estimate strains and stresses from changes in the lattice parameter. Generally, the residual stresses are tensile stresses in three directions in most cases, but the stress level is different. The residual stresses are largest in the transverse direction, followed by a medium level in oscillation direction, and lowest in the axial direction. This difference is attributed to the thermal gradient and welding pressure according to Bhamji et al.[7]

Daymond and Bonner[30] measured the strain in LFW dissimilar Ti64 joints (MDG10051R and MDG10050) by pulsed neutron diffraction. It was estimated that within the welding plane, strains were large and tensile in the HAZ; in the region around the weld strains became compressive, and in the far field, the tensile strains reduced to zero slowly. However, the axial strains inside the HAZ and weld line were compressive, but they became tensile in the region outside the HAZ. A moderate component of hydrostatic tensile stress existed in the WCZ, superimposed with a biaxial tensile stress in the welding interface.

Romero et al.[26] and Turner et al.[29] found that the effect of forging pressure on the residual stresses development in Ti64 linear friction welds was significant. Figure 9.8[26] shows the influence of the friction pressure on the residual

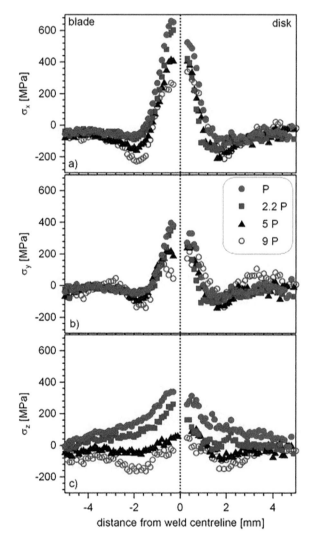

**FIGURE 9.8**

Influence of the friction pressure on the residual stress components (*x*-direction: transverse direction, *y*-direction: oscillation direction, *z*-direction: axial direction).[26] (With kind permission from Elsevier.)

stresses distribution for all three principal directions for a range of friction pressures. As the friction pressure increases, the residual stresses reduce significantly. By increasing the friction pressure, welds of lower temperature are produced compared to low-pressure welds. Power input may be increased by increasing the friction pressure, but welding time is reduced and material flow is accelerated during the process with high pressure. These cause a drop in the total weld energy and a reduction in the total heat load. It can also be seen that the highest tensile stresses develop near the weld line in the x-direction. The reason for this is that the distance from the WCZ to the outside surface was longer in the x-direction, relative to that in the y-direction, and therefore it was likely that the thermal gradient and thermal mismatch will be larger in the x-direction. In addition, Turner et al.[29] also developed a numerical model to predict the residual stresses in LFW Ti64 joints, and the results were comparable to the experiments to some extent, which provided an easy way for future stress prediction if the model could be elaborately modified.

PWHT has been shown to significantly reduce the residual stresses in linear friction welds. Frankel et al.[31] reported the influence of PWHT on residual stresses in Ti64 and Ti-6Al-2Sn-4Zr-2Mo (Ti6242) joints by high-energy synchrotron X-ray diffraction. Figure 9.9 shows the change of residual stresses for the two alloys after different PWHTs.[31] As expected, after a heat treatment (100°C) below the conventional aging temperature of Ti64 (PWHT1), the value of tensile peak stress was about 500 MPa in the transverse direction, meaning that the effect of PWHT1 was useless. The conventional heat treatment (PWHT2) for Ti64 was shown to reduce the stresses in Ti64 to below 50 MPa. In a similar case, residual stresses of about 450 MPa remained in Ti6242 weld after PWHT2 (see Figure 9.9c,e). Considering the excellent high-temperature capabilities of the near $\alpha$ alloy, this result was acceptable. As for PWHT3, a higher temperature (100°C over conventional heat treatment temperature) was used. The stresses were released below 100 MPa (Figure 9.9f). It can be seen that the PWHTs are effective in relieving residual stresses. Preuss et al.[28] also investigated the effect of PWHT on the residual stresses of LFW Ti-4Al-2Sn-4Mo-0.5Si (IMI550) joints. When increasing the PWHT temperature by 100°C or 150°C, the residual stresses can be reduced to a moderate level. All conditions (as-welded and PWHT) displayed a significant hydrostatic tensile stress field component, which was generally difficult to relieve.

### 9.3.5 Hardness Profiles of Joints

Hardness measurements along the joints have been made for a number of titanium alloys including Ti64,[5,26,31] Ti17,[10] Ti6242,[32] Ti6246,[21,33] and Ti5553,[11] which are useful in the interpretation of welding microstructure and mechanical properties. The various alloy types and welding conditions will lead to the different shape of the hardness profile. Hardness shows a W-shaped profile with the highest values occurring in the WCZ in most cases. The phase

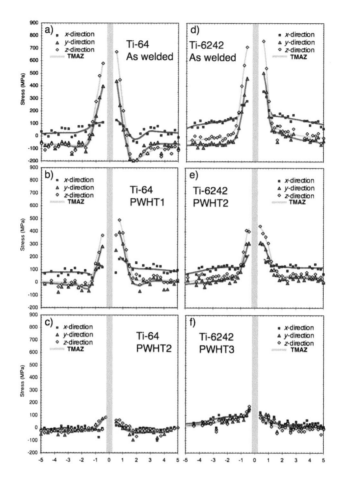

**FIGURE 9.9**
Residual stress in three principal directions (*x*-direction: oscillation direction, *y*-direction: transverse direction, *z*-direction: axial direction) for (a–c) Ti64 and (d–f) Ti6242 in as-welded and PWHT conditions.[31] (With kind permission from T&F.)

transformation and grain refinement are the main reason for this high value in WCZ.[5] In all cases, hardness drops rapidly to values lower than the PM in the region adjacent to the WCZ/TMAZ. The low value of microhardness in TMAZ can be attributed to the mixing of the banded microstructure during deformation of the interface material in the TMAZ.[5]

However, the different microhardness profile in LFW Ti17 joints was found by Li et al.[10] where the hardness in the WCZ and TMAZ was slightly lower than that in PM, with the average hardness in TMAZ being lower than that in WCZ. The profile of the hardness curve in the TMAZ was concave with the highest value near the WCZ. The microhardness value in WCZ was the highest of these joints exhibited at the WCZ, and it gradually became lower to that in PM (Figure 9.10). The nonuniform deformation and microstructural

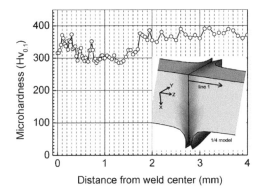

**FIGURE 9.10**
Microhardness distribution from the WCZ to the parent Ti17.[10] (With kind permission from John Wiley and Sons.)

changes in the TMAZ cause this hardness profile. In addition, the presence of multiple $\beta$ grains in WCZ may cause the relatively low hardness in this region due to the fact that the $\beta$ phase is usually softer than the $\alpha$ phase.

In addition, Guo et al.[21] found an opposite hardness profile in LFW Ti6246 joints as shown in Figure 9.11. Their explanation was the co-existing of $\beta$ phase and the soft $\alpha''$ martensite in the as-welded joints. After PWHT, the WCZ consisted of the equilibrium $\beta$ and $\alpha$ phases and the hardness in this region was increased. Dalgaard et al.[11] also reported the similar hardness profile in as-welded Ti-5553 joints. They suggested that the absence of hard globular $\alpha$ phase in WCZ caused the decrease of microhardness.

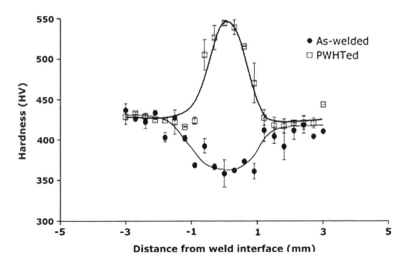

**FIGURE 9.11**
Microhardness profile across the weld interface at as-weld and post-weld heat treated conditions obtained from the LFW Ti6246 weld.[21] (With kind permission from Elsevier.)

The effect of welding parameters on microhardness was investigated by Romero et al.[26] and Wanjara and Jahazi.[5] Romero et al.[26] pointed out that the width of high-hardness region was decreased with increasing friction pressure, while changes in pressure had a negligible effect on the peak values of the hardness of the WCZ. Wanjara and Jahazi[5] found that the hardness value in the TMAZ and the size of TMAZ decreased with increasing frequency, which can be explained by the varying material extrusion for different process parameters of LFW. The heat input was increased and materials could be plasticized more easily and more rapidly with increasing frequency. Amplitude and friction pressure affected hardness in a similar fashion to frequency. But the effect of axial shortening change on the minimum hardness value in the TMAZ was negligible.

Mateo et al.[34] performed single, double, and triple welds to simulate repairs with one or two replacements of the blade in the original position. The hardness profile measured of the single weld (Figure 9.12a) was nearly identical to other studies of Ti64.[5,26] Two similar peaks were found in the hardness profile of the double weld (Figure 9.12b), but the highest values were lower in value to that of the single weld. A similar distribution was observed (Figure 9.12c) in the case of triple weld. It can be concluded that multiple welding processes will reduce the peak value of hardness profile, and it can be attributed to the removal of residual stresses caused by heat history during multiple welding processes which can be regarded as a special PWHT.

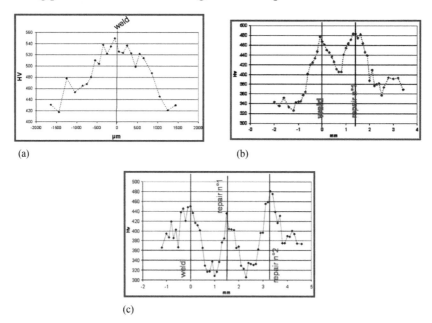

(a)

(b)

(c)

**FIGURE 9.12**
Hardness profile of (a) single weld, (b) double weld, and (c) triple weld.[34] (With kind permission from T&F.)

### 9.3.6 Tensile Properties of Joints

Although the materials were different, the similar behavior was observed where the fracture occurred in the PM far away from the weld if there are not any defects in the interface, indicating that the strength of the joints exceeds the strength of the PM during tensile testing.[33–35]

Wanjara and Jahazi[5] focused on the influence of welding parameters on tensile mechanical properties of Ti64. They found that the effects of frequency and amplitude on the joint quality were most significant, and the effects pressure and axial shortening of on the weld quality were secondary. In addition, they pointed out that the fracture location of LFW joints during tensile testing was influenced by the processing conditions. For processing conditions which produced sound welds at PI (power input) values higher than 2.4 kW, the fracture was observed to occur in the PM. For PI values lower than 2.4 kW caused by low frequency or low amplitude, and when defects (such as oxides) existed in the welding interface, the tensile fracture occurred in the region of the TMAZ.

Li et al.[10,20] also measured the tensile strength of Ti17 where joints exhibited similar strength values to PM. One of the three samples fractured near the WCZ and the resultant microstructure had a significant impact on the fracture features. It was concluded that the specific properties of the joints were dependent on the particular nonequilibrium heating and cooling conditions during the welding process. In addition, when the samples were heat treated at a temperature between 610°C and 670°C,[20] the tensile strength of joints was almost the same as the PM however with lesser elongation.

Dalgaard et al.[11] found that the tensile strength of LFW near-$\beta$ Ti-5553 joints was a little lower than that of the base material, while the elongation was much lower. The reduction in tensile strength of the welded samples was explained by the presence of a softened region in the weldment with a microstructure depleted of $\alpha$ phase.

### 9.3.7 Impact Toughness, Fracture Toughness, Fatigue Properties, and Torsional Strength of Joints

Although impact toughness, fracture toughness, fatigue properties, and torsional strength are mechanical properties which are essential in the design of high-performance structures, there are a limited number of published studies on these.

Ma et al.[13] used Mesnager U-notch specimens to analyze the impact toughness and fracture features of LFW Ti64 alloy joints. All of the specimens fractured in the PM. In most cases, the initiation of the crack occurred in the WCZ. Then, the creak propagated through the TMAZ into the PM quickly, which indicated that impact toughness of the weld was higher than that of the PM. Ma et al.[20] also investigated the impact toughness of linear friction welded Ti17 joints. It was found that the mean impact toughness of the samples after PWHT at 530°C was slightly superior to that of the as-welded joint. It reached 80.9% of that of the parent metal after treatment at 610°C. When the treatment temperature is 670°C, the average impact toughness could be 93.4% of the parent metal. Fracture analysis after mechanical testing shows that the fracture behavior of the samples changes from brittle failure to ductile one with the increase in PWHT temperature.

The fracture toughness of Ti6246 was measured by Corzo et al.,[33] and they found that the values were lower than the PM. In one case, both the initiation and propagation of crack were developed along the weld line, which was attributed to the very fine microstructure in WCZ where the hardness value was high, and it was usually associated with reduced fracture toughness. In other experiments presented in this paper, the crack propagated toward the HAZ, with the fracture toughness values correlating to the HAZ toughness. Tao et al.[36] measured the fracture toughness of LFW Ti64-Ti17 joints and found that the WCZ and TMAZ on Ti17 side were the weak zones in the joints with low fracture toughness values, and these low fracture toughness values can be improved effectively by PWHT, but the effect of PWHT on the improvement of fracture toughness on Ti64-side TMAZ was negligible. However, for the Ti17 joints, Ji and Wu[37] found that the minimum fracture toughness was in the weld center in as-welded condition and PWHT condition.

Mateo et al.[34] tested the fatigue life response of a repaired Ti64 part. It was observed that after high cycle fatigue, all of the joints fractured in the PM rather than WCZ. Same results were also found by Wen et al.,[38] Wang et al.,[39] and Stinville et al.[23] For the dissimilar titanium joints, Wen et al.[40] investigated the fatigue properties of Ti64–Ti11 joints and found fatigue failure occurred on the Ti64 side and was far from the weld line, but Zhao et al.[41] found the fragment of the joint locates at the WCZ of Ti17 side in the Ti11–Ti17 joints, which can be ascribed to the coarsened beta phase on the WCZ of the Ti17 side.

Suleimanova et al.[42] presented the torsional strength results of linear friction welded VT6 joints. The torsional strength value of the weld was superior to the PM due to the test failures in the PM at a rotation degree of 110°.

## 9.4 Summary

LFW has received increasing attention in recent years. Some efforts have provided invaluable insight into the nature of the LFW process. A review of the published literature about LFW of titanium alloys has been conducted. From this survey, the following conclusions can be drawn: The LFW microstructure is characterized by a WCZ where grain refinement occurs due to recrystallization accompanied with metallurgical phase transformation; the

TMAZ is close to WCZ and consists of deformed and/or only partly recrystallized grains. One or more strong texture can be observed in WCZ under the effect of phase transitions and deformation. The texture distribution is affected by welding parameters and specimen size, but the formation mechanism is not clear so far. Significant residual stresses have been reported in LFW joints due to the thermal mismatch caused by various cooling rates at different positions across the weld. Changing the welding parameters and PWHTs can reduce the residual stresses partially. The mechanical properties of welds tend to be superior in the WCZ due to the refined microstructure. Some opposite results can be attributed to phase transitions or abnormal microstructure formation. More work is needed to better understand the interrelationship between the welding parameters and the distributions of welding temperature and effective method to obtain the range of temperatures by changing welding parameters. It could be significant to improve our understanding of the process and to optimize welding parameters for LFW efficiently.

# References

1. Vairis, A., and M. Frost. 1998. High frequency linear friction welding of a titanium alloy. *Wear* 217: 117–131.
2. Vairis, A., and M. Frost. 1999. On the extrusion stage of linear friction welding of Ti 6Al 4V. *Mater Sci Eng A Struct Mater* 271: 477–484.
3. Vairis, A., and M. Frost. 2000. Modelling the linear friction welding of titanium blocks. *Mater Sci Eng A Struct Mater* 292: 8–17.
4. Bhamji, I., M. Preuss, P. L. Threadgill, et al. 2010. Linear friction welding of AISI 316L stainless steel. *Mater Sci Eng A Struct Mater* 528: 680–690.
5. Wanjara, P., and M. Jahazi. 2005. Linear friction welding of Ti-6Al-4V: Processing, microstructure, and mechanical-property inter-relationships. *Metall Mater Trans A* 36(8): 2149–2164.
6. Ma, T. J., T. Chen, W. Y. Li, et al. 2011. Formation mechanism of linear friction welded Ti-6Al-4V alloy joint based on microstructure observation. *Mater Charact* 62: 130–135.
7. Bhamji, I., M. Preuss, P. L. Threadgill, et al. 2011. Solid state joining of metals by linear friction welding: A literature review. *Mater Sci Technol* 27(1): 2–12.
8. Li, W., A. Vairis, M. Preuss, et al. 2016. Linear and rotary friction welding review. *Int Mater Rev* 61(2): 71–100.
9. Lang, B., T. C. Zhang, X. H. Li, et al. 2010. Microstructural evolution of a TC11 titanium alloy during linear friction welding. *J Mater Sci* 45: 6218–6224.
10. Li, W. Y., T. J. Ma, and S. Q. Yang. 2010. Microstructure evolution and mechanical properties of linear friction welded Ti-5Al-2Sn-2Zr-4Mo-4Cr (Ti17) titanium alloy joints. *Adv Eng Mater* 12(1–2): 35–43.
11. Dalgaard, E., P. Wanjara, J. Gholipour, et al. 2012. Linear friction welding of a near-β titanium alloy. *Acta Mater* 60: 770–780.

12. Karadge, M., M. Preuss, C. Lovell, et al. 2007. Texture development in Ti–6Al–4V linear friction welds. *Mater Sci Eng A Struct Mater* 459: 182–191.
13. Ma, T. J., W. Y. Li, and S. Y. Yang. 2009. Impact toughness and fracture analysis of linear friction welded Ti-6Al-4V alloy joints. *Mater Des* 30: 2128–2132.
14. Sun, D., Z. Ren, Z. Zhou, et al. 2000. Microstructural features of friction-welded Ti-6Al-4V joint. *J Mater Sci Technol* 16: 59–62.
15. Maio, L., M. Liberini, D. Campanella, et al. 2017. Infrared thermography for monitoring heat generation in a linear friction welding process of Ti6Al4V alloy. *Infrared Phys Technol* 81: 325–338.
16. Guo, Y., M. M. Attallah, Y. Chiu, et al. 2017. Spatial variation of microtexture in linear friction welded Ti-6Al-4V. *Mater Charact* 127: 342–347.
17. Ahmed, T., and H. J. Rack. 1998. Phase transformations during cooling in $\alpha+\beta$ titanium alloys. *Mater Sci Eng A Struct Mater* 243(1): 206–211.
18. McAndrew, A. R., P. A. Colegrove, C. Bühr, et al. 2018. A literature review of Ti-6Al-4V linear friction welding. *Prog Mater Sci* 92: 225–257.
19. Wang, X. Y., W. Y. Li, T. J. Ma, et al. 2017. Characterisation studies of linear friction welded titanium joints. *Mater Des* 116: 115–126.
20. Ma, T. J., W. Y. Li, B. Zhong, et al. 2012. Effect of post-weld heat treatment on microstructure and property of linear friction welded Ti17 titanium alloy joint. *Sci Technol Weld Join* 17(3): 180–185.
21. Guo, Y., N. T. Jung, Y. L. Chiu, et al. 2013. Microstructure and microhardness of Ti6246 linear friction weld. *Mater Sci Eng A Struct Mater* 562: 17–24.
22. Schröder, F., R. M. Ward, A. R. Walpole, et al. 2015. Linear friction welding of Ti6Al4V: Experiments and modelling. *Mater Sci Technol* 31(3): 372–384.
23. Stinville, J. C., F. Bridier, D. Ponsen, et al. 2015. High and low cycle fatigue behavior of linear friction welded Ti–6Al–4V. *Int J Fatigue* 70: 278–288.
24. Li, W. Y., H. Wu, T. J. Ma, et al. 2012. Influence of parent metal microstructure and post-weld heat treatment on microstructure and mechanical properties of linear friction welded Ti-6Al-4V joint. *Adv Eng Mater* 14(5): 312–318.
25. Li, W., J. Suo, T. Ma, et al. 2014. Abnormal microstructure in the weld zone of linear friction welded Ti–6.5 Al–3.5 Mo–1.5 Zr–0.3 Si titanium alloy joint and its influence on joint properties. *Mater Sci Eng A Struct Mater* 599: 38–45.
26. Romero, J., M. M. Attallah, M. Preuss, et al. 2009. Effect of the forging pressure on the microstructure and residual stress development in Ti-6Al-4V linear friction welds. *Acta Mater* 57: 5582–5592.
27. Gey, N., and M. Humbert. 2002. Characterization of the variant selection occurring during the $\alpha \rightarrow \beta \rightarrow \alpha$ phase transformations of a cold rolled titanium sheet. *Acta Mater* 50(2): 277–287.
28. Preuss, M., J. Quinta Da Fonseca, A. Steuwer, et al. 2004. Residual stresses in linear friction welded IMI550. *J Neutron Res* 12(1–3): 165–173.
29. Turner, R., R. M. Ward, R. March, et al. 2012. The magnitude and origin of residual stress in Ti-6Al-4V linear friction welds: An investigation by validated numerical modeling. *Metall Mater Trans* 43: 186–197.
30. Daymond, M. R., and N. W. Bonner. 2003. Measurement of strain in a titanium linear friction weld by neutron diffraction. *Physica B* 325: 130–137.
31. Frankel, P., M. Preuss, A. Steuwer, et al. 2009. Comparison of residual stresses in Ti-6Al-4V and Ti-6Al-2Sn-4Zr-2Mo linear friction welds. *Mater Sci Technol* 25(5): 640–650.

32. Baeslack III, W. A., T. F. Broderick, M. Juhas, et al. 1994. Characterization of solid-phase welds between Ti-6Al-2Sn-4Zr-2Mo-0.1Si and Ti-13.5Al-21.5Nb titanium aluminide. *Mater Charact* 33: 357–367.
33. Corzo, M., Y. Torres, M. Anglada, et al. 2007. Fracture behaviour of linear friction welds in titanium alloys. *Anales de la Mecánica de Fractura* 1: 75–80.
34. Mateo, A., M. Corzo, M. Anglada, et al. 2009. Welding repair by linear friction in titanium alloys. *Mater Sci Technol* 25(7): 905–913.
35. Li, W. Y., T. J. Ma, Y. Zhang, et al. 2008. Microstructure characterization and mechanical properties of linear friction welding Ti-6Al-4V alloy. *Adv Eng Mater* 10(1–2): 89–92.
36. Tao, B. H., Q. Li, Y. H. Zhang, et al. 2015. Effects of post-weld heat treatment on fracture toughness of linear friction welded joint for dissimilar titanium alloys. *Mater Sci Eng A Struct Mater* 634: 141–146.
37. Ji, Y., S. Wu. 2014. Study on microstructure and mechanical behavior of dissimilar Ti17 friction welds. *Mater Sci Eng A Struct Mater* 596: 32–40.
38. Wen, G. D., T. J. Ma, W. Y. Li, et al. 2014. Cyclic deformation behavior of linear friction welded Ti6Al4V joints. *Mater Sci Eng A Struct Mater* 597: 408–414.
39. Wang, S. Q., T. J. Ma, W. Y. Li, et al. 2017. Microstructure and fatigue properties of linear friction welded TC4 titanium alloy joints. *Sci Technol Weld Join* 22(3): 177–181.
40. Wen, G. D., T. J. Ma, W. Y. Li, et al. 2014. Strain-controlled fatigue properties of linear friction welded dissimilar joints between Ti–6Al–4V and Ti–6.5 Al–3.5 Mo–1.5 Zr–0.3 Si alloys. *Mater Sci Eng A Struct Mater* 612: 80–88.
41. Zhao, P., L. Fu, and H. Chen. 2016. Low cycle fatigue properties of linear friction welded joint of TC11 and TC17 titanium alloys. *J Alloys Compd* 675: 248–256.
42. Suleimanova, G., R. Kabirov, M. Karavaeva, et al. 2015. Investigation of torsional strength of the VT6 weld joint produced by linear friction welding. *Russ Phys J* 58(6): 815–821.

# 10

# *Joining of Dissimilar Materials Using Friction Welding*

**Udaya Bhat K.**

*National Institute of Technology Karnataka*

**Suma Bhat**

*Srinivasa School of Engineering*

## CONTENTS

10.1 Introduction ....................................................................................... 212
10.2 Dissimilar Materials Joint Design .................................................. 212
10.3 Joining Processes for Dissimilar Materials .................................. 213
    10.3.1 Arc-Based Fusion Welding Processes ............................... 213
    10.3.2 Non-Arc-Based Fusion Welding Processes ...................... 213
    10.3.3 Solid-State Processes ........................................................... 214
    10.3.4 Brazing, Soldering, and Adhesive Bonding .................... 214
    10.3.5 Mechanical Fastening .......................................................... 214
10.4 Friction Welding ............................................................................... 214
    10.4.1 Basic Steps Involved in Friction Welding ........................ 214
    10.4.2 Rotary Friction Welding ...................................................... 216
    10.4.3 Orbital Friction Welding ..................................................... 218
    10.4.4 Mechanism of Friction Welding ......................................... 218
        10.4.4.1 Friction Stage ....................................................... 218
        10.4.4.2 Forging Stage ....................................................... 219
    10.4.5 Friction Welding Process Parameters ................................ 220
        10.4.5.1 Speed ..................................................................... 220
        10.4.5.2 Pressure ................................................................. 220
        10.4.5.3 Heating Time ........................................................ 221
    10.4.6 Heat Generation in Friction Welding ................................ 221
    10.4.7 HAZ during Friction Welding ............................................ 221
10.5 Friction Welding of Pure Dissimilar Metal Combinations .......... 222
    10.5.1 Deformation Behavior .......................................................... 223
    10.5.2 Burn-Off and Reaction Layer Thickness ........................... 223
    10.5.3 Heat-Affected Zone .............................................................. 224
    10.5.4 Mixing at the Interface ......................................................... 224

10.6    Use of Friction Welding for Dissimilar Materials Joining...............224
        10.6.1  Use of Friction Welding for Dissimilar Joints Involving
                Al Alloys...................................................................225
        10.6.2  Welding of Dissimilar Steels.....................................227
        10.6.3  Friction Welding of Steels with Other Metals......................228
        10.6.4  Copper and Copper Alloys.........................................229
10.7    Friction Welding of Ceramics.............................................229
10.8    Friction Welding of Polymers............................................231
10.9    Summary..................................................................231
References.......................................................................232

## 10.1 Introduction

Design engineers are always looking to create new parts and structures with tailored properties. Sometimes a part needs to be lightweight, on the one end, and corrosion resistant on the other end. In some other situation, a component needs to have high corrosion resistance on the one end, and simultaneously it needs to have high-temperature resistance on the other. Possibility of joining dissimilar metals and alloys with different sets of physical, mechanical, thermal, and corrosion resistance behaviors enables new designs in manufacturing, automotive, energy production, medical and consumer applications (Meshram, Mohandas, and Reddy 2007, 330; Ji, Wu, and Zhao 2016, 197; Teker 2013, 303; Alves et al. 2012, 20; Tenaglia 2012; Gould 2011; Ferranate and Pigoretti 2002, 2825; Fabritsiev et al. 1998, 2030).

Dissimilar joints are not limited to metallic materials alone. Joining of thermoplastic matrix materials to thermosetting composites is a highly relevant concept offering the highest flexibility to polymer design engineer (Ageorges and Ye 2001, 1603). Similarly, metal-to-polymer joints offer the benefit of strength and ductility of the metal with lightweight and corrosion resistance of the polymer (Balle, Wagner, and Eifler 2009, 1). Joining of lightweight dissimilar materials, like metals and polymers, is important in the design of hybrid structures (Singh et al. 2016, 77; Nagatsuka et al. 2015, 82). With the increased thrust on the development of the lightweight structures in aerospace and automotive sectors, possibility of getting a good dissimilar joint is highly relevant (Kah et al. 2014, 152; Sakiyama et al. 2013).

## 10.2 Dissimilar Materials Joint Design

Successful dissimilar materials joints are often important for the technical performance and commercial success of a component having dissimilar materials in it. At the same time, joining of dissimilar materials is more difficult compared to joining of similar materials. However, with the development of new joining processes and procedures, many combinations of dissimilar materials could be successfully joined (Tenaglia 2012; Gould 2011; Kim et al. 2010; Balle, Wagner, and Eifler 2009, 1).

There are many factors that affect the design of a dissimilar material joint. Some of them are listed below:

   i. Joint design and materials thicknesses

   ii. Mismatch in the coefficient of thermal expansion

   iii. Fixtures, constraint effects, and stress development

   iv. Metallurgical compatibility and formation of the brittle phases

   v. Phase transformations during cooling and the effect of cooling rate on it

   vi. Requirements of pre- and post-heat treatment steps to minimize the stresses

   vii. Effect of environment on the weldment and nearby region

   viii. Post-weld life environment (Tenaglia 2012)

   ix. Relative dimensions of two pieces, which will act as heat sinks

## 10.3 Joining Processes for Dissimilar Materials

Almost all welding processes have been explored for the possibility of joining dissimilar materials. Also, note that the other joining processes, like riveting and mechanical fitting, are also explored. Each process has its own processing parameters and procedure to be used to get the best joint performance. Each process has its own set of advantages and disadvantages. Broadly, these processes could be grouped as follows.

### 10.3.1 Arc-Based Fusion Welding Processes

This group consists of shielded metal arc welding (SMAW), gas tungsten arc welding (GTAW), gas metal arc welding (GMAW), plasma arc welding (PAW), etc.

### 10.3.2 Non-Arc-Based Fusion Welding Processes

This group consists of laser welding, resistance spot welding, resistance seam cladding, electron beam (EB) welding, etc.

### 10.3.3 Solid-State Processes

This group consists of friction welding, friction stir welding, friction spot welding, ultrasonic welding, diffusion welding, explosive welding, roll cladding, etc.

### 10.3.4 Brazing, Soldering, and Adhesive Bonding

This group consists of allied welding processes such as brazing, soldering, adhesive bonding, etc. (Tenaglia 2012; Kah et al. 2014, 152; Filho and dos Satos 2009, 1461).

### 10.3.5 Mechanical Fastening

Since the focus of this chapter is on the solid-state processes based on the friction energy, friction welding would be elaborated in subsequent pages, with the aim of analyzing its effects on metallic materials. This group consists use of rivets, use of nuts and bolts, etc. (Kah et al. 2014; Filho and dos Satos 2009, 1461).

## 10.4 Friction Welding

Friction welding is a solid-state joining method, which is one of the most economical and highly productive methods for joining similar and dissimilar metals. It is widely used in automotive and aerospace component fabrication. It is often the only practical alternatives available in the joining domain available to overcome the difficulties encountered during the joining of the materials with widely varying physical properties (Uday et al. 2010, 534; James and Sudhish 2016, 1191; Winiezenko and Kaczorowski 2012, 444). Main advantages of friction welding are high materials saving, low production time, and the possibility of welding of dissimilar metals or alloys (Teker 2013, 303; Winiezenko and Kaczorowski 2012, 444).

### 10.4.1 Basic Steps Involved in Friction Welding

Some of the basic steps involved in friction welding are explained below. The procedure explained here is for friction welding of two rods of similar materials with the same diameter. Practically, both conditions are hardly met. Problem is more challenging when both differ considerably.

Conventionally, term rotary friction welding is used when both parts are of the same diameter and they need to be welded along the geometrical axis (Figure 10.1).

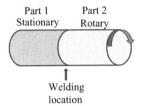

**FIGURE 10.1**
Configuration of the parts during rotary friction welding.

Here, the friction welding is carried out by moving one component with respect to another component along a common interface, while simultaneously applying a normal force along the common axis of the rods to be welded. The friction heating generated at the interface due to the relative motion of the rods and normal force at the interface softens both or one of the components (depending on whether the rods are of similar or dissimilar rods). When the material is sufficiently plasticized using an additional force (forge force), the plasticized material is expelled out from the common interface. This leaves clean material on both rods in contact. With the application of the appropriate brake, the relative motion is suddenly stopped. The joint is facilitated with the application of forge force in time (Figure 10.2). The weld is allowed to cool. The weld forms when the rods are in solid state and no molten material is generated (Uday et al. 2010, 534; Lienert et al. 2012).

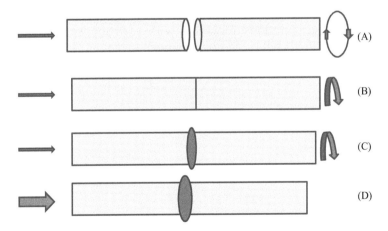

**FIGURE 10.2**
Schematic presentation of steps in the friction welding. A: rotary part is being brought toward a stationary part. B: Rotary part is in the rotation against a stationary part, with the simultaneous application of the friction force. C: A small region near the interface is getting heated (indicated by dark color). D: Rotation is stopped, and the forging force is being applied (flash is indicated by the dark color). The thickness of the arrow scales with the magnitude of the force.

### 10.4.2 Rotary Friction Welding

Rotary friction welding accounts for most of the friction welding activities (Nicholas 2003, 2). There are two variants of this process, namely, direct drive and inertia drive. This classification is based on the energy source. The direct drive uses energy required to weld from an infinite source, whereas inertia drive uses energy stored in a flywheel. Direct drive process is a well-established process (Nguyen and Weckman 2006, 275), which can be used for various types of steels and metallurgically difficult to weld systems, like dissimilar combinations and super alloys (Kurt, Uygur, and Paylasan 2011; Reddy, Rao, and Mohandas 2008, 619).

In direct drive friction welding, one workpiece is rotated at a predetermined speed, using a motor driven unit and the other is restrained from the rotation (Nicholas 2003, 2). The parts to be welded are brought together, and a friction force is applied. Due to the combined action of the rotation and the friction force, heat is generated at the faying surfaces (weld interface). This situation is continued for a predetermined time or for a preset upset value. The rotational torque is discontinued, and rotation of the part is stopped either by its own resistance or through a braking force. Figure 10.3 shows a schematic variation of various parameters as a function of time.

Inertial drive welding is another popular variant of rotary friction welding, and is very useful for joining axisymmetrical parts (Li et al. 2007, 1408; El-Hardek 2014, 2346). Here one part is restrained from rotating (O'Brien 1991; Daus et al. 2007, 1424; Mortensen et al. 2001, 268), while the second part is connected to a flywheel, which is accelerated to a predetermined speed using a drive motor. The flywheel is used to store the required energy. Once the flywheel is accelerated to the required speed, the drive motor is disengaged and the parts are forced together with a friction welding force. The kinetic energy stored in the rotating flywheel is dissipated so that the heat at the interface causes parts to be heated to plasticizing temperature. Later,

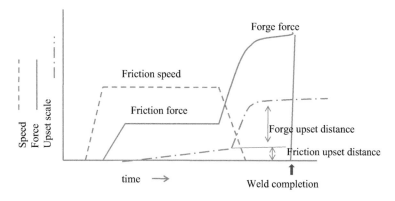

**FIGURE 10.3**
Parameters in direct drive friction welding, sketch not to the scale.

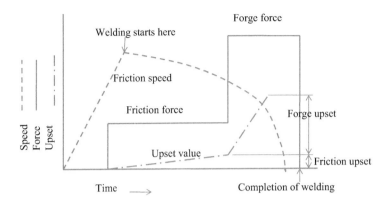

**FIGURE 10.4**
Parameters in inertia friction welding process, sketch not to the scale.

a forging force is applied as the rotation stops (Nicholas 2003, 2; Sahin 2007, 2244; Sahin 2009a, 319). The forge force is retained for certain time even after rotation stops. Figure 10.4 presents various parameters of inertia friction welding. Rotary friction welding has the major limitation that it is not suitable for joining noncircular sections. Also, the rate of heat generation is not uniform across the cross section leading to nonuniform heat-affected zones (HAZs) with an appropriate set of process parameters, this needs to be minimized.

The industry has developed different varieties based on the nature of the components to be joined. Some of them are linear friction welding and orbital friction welding. In the case of the former, the components to be welded are made to move in a reciprocating fashion with a small displacement relative to each other, under friction pressure. In contrast, in orbital friction welding, the center of one component is made to move relative to the other in a circle to generate frictional heat. The two components to be joined are rotated with the same angular speed along their longitudinal axis, with a small offset within them. Just before the moment of application of forging pressure, the motion of the component is stopped, and the parts are correctly aligned to form a weld (Uday 2010, 534; Maalekian 2007, 738; O'Brien 1991; Maalekian et al. 2008, 2843; D'Alvise, Massoni, and Walloe 2002, 387).

In rotary friction welding, the outer surfaces generate and experience high heat. This limitation could be avoided by using either linear or orbital friction welding. They can weld noncircular sections as well as produce uniform heat generation at the interface (Figure 10.5). Again heat generation in orbital welding is more uniform compared to that in linear friction welding. It is attributed to the uniform relative velocity between the two components over the whole area (El-Hardek 2014, 2346; Alves et al. 2012, 20; Uday 2010, 534; Maalekian et al. 2008, 2843). It may be noted that in linear friction welding, the outer surfaces generate lesser heat. In orbital welding, combinations of

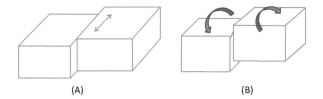

**FIGURE 10.5**
Contacts and faying surfaces in A: Linear and B: Orbital friction welding geometries.

linear and rotary movements are used to produce and distribute uniform heat and to get uniform temperature.

### 10.4.3 Orbital Friction Welding

This is suitable for the joining of noncircular cross-sectional parts also (Maalekian 2007, 738). In this variant, neither of the workpieces rotate about their central axis; instead, they move in an orbital fashion to provide uniform tangential velocity over the entire interface area. As soon as the orbital motion is stopped, parts are realigned quickly. Practically, it is possible to provide high-quality welds, and it is attributed to the constancy of the relative velocity at the rubbing surface. It generates a uniform rate of heat generation at the interface and hence uniform HAZ. Cross section can be of any shape. Compared to rotary friction welding, the amount of literature available is limited.

### 10.4.4 Mechanism of Friction Welding

The welding cycle in Frcition Welding (FW) can be broadly divided into two stages: friction stage and upsetting stage. In the first stage, welding heat is generated, and in the second stage, the weld is consolidated and the joint is cooled.

#### 10.4.4.1 Friction Stage

Assuming that two parts are identical in dimensions following mechanical processes take place. When two parts are brought in contact with a force, adhesion occurs at various points of contact. Since contact is at a few sites and the sites are extremely small in nature, the local pressure would be extremely high. Depending on the chemistry of the parts, contamination, etc., the adhesion may be stronger than the cohesion in the metal on either side. As the rod rotates, shearing takes place at the point of contact and the metal fragments are transferred from one side to the other. As rotation progresses, both torque and interfacial temperature increases. Similarly, the amount of fragments increases and a continuous layer of plasticized metal forms at the interface. The incipient liquid film may form at this point. Also, the value of the torque peaks and will remain reasonably constant as metal is getting heated and axial shortening occurs due to the ejection of the plasticized metal laterally (O'Brien 1991; Kimura et al. 2003, 384; Lienert et al. 2012).

### 10.4.4.2 Forging Stage

Forging stage starts immediately after the end of the heating stage (refer Figure 10.3). Here upsetting of the rods take place to cause axial shortening. Upset is visible in the form of a flash (refer the schematic in Figure 10.6). For similar metals with similar diameter, flash on both sides will be similar. In the case of dissimilar materials combination, the magnitude of the flash would be asymmetric (Figure 10.7). At this stage, bond at the interface occurs and the joint cools from its maximum temperature.

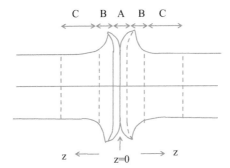

**FIGURE 10.6**
Sketch indicating different zones during friction welding of similar metallic materials. A: fully plasticized zone, B: Partially deformed zone, C: Undeformed region. Boundary lines for the zones are imaginary lines drawn.

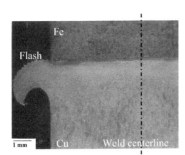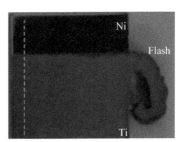

**FIGURE 10.7**
Macrographs indicating the deformation of the part having lower strength. In Ti-Ni combination, Ti deforms, and in Cu-Fe combination, Cu deforms.

The bonding mechanism is more complex in the case of dissimilar metals joining. The formation of the bond depends on a large number of parameters like surface energy, crystal structure, mutual solubility, and the possibility of formation of the intermetallic compounds (IMCs). Some amount of alloying is also expected to occur in a narrow region at the interface, as a result of mechanical mixing and diffusion. The property of this layer has a significant effect on the dissimilar metal joint performance. Mechanical mixing, chemical reaction, physical interlocking, etc. affect the joint properties. Due to these concerns, predicting the weld quality for many dissimilar metal combinations is practically difficult. Suitability of a particular combination should be established after a series of experiments suitably designed based on the performance requirements (Kimura et al. 2003, 384).

### 10.4.5 Friction Welding Process Parameters

Important process parameters affecting the performance of the weld joint, either individually or in combination, are speed, pressure, and heating time (Lienert et al. 2012). In linear and orbital friction welding, the amplitude is another variable along with the previously mentioned parameters.

#### 10.4.5.1 Speed

The rotational speed is related to the material to be welded and the diameter of the weld at the interface. Increasing rotational speed might lead to greater frictional heat at the interface; its consequence is increased softening of the material and greater recrystallization. In the case of certain materials, it can lead to thicker intermetallic layer (IML) formation. An important aspect to be noted is that depending on the physical and mechanical properties of the materials, different ranges of rotational speeds produce different effects on the quality of the weld. Therefore, it is essential to use appropriate speeds to minimize the detrimental effect and produce joints of good quality (Jamaludin, Keat, and Ahmad 2004, 85). It is more so in the case of a dissimilar metals combination.

#### 10.4.5.2 Pressure

Pressure is also an important parameter controlling the temperature gradient in the weld zone, along with the required drive power, and the axial shortening. The value of the pressure depends on the metals to be joined and their joint geometry. Heating (friction) pressure needs to be high enough to keep faying surfaces in intimate contact to avoid oxidation (Mousavi and Kelishami 2008, 178s). For a given spindle speed, lower pressure limits the heating effect and hence little or no axial shortening. Similarly, the joint quality is improved in many metals by applying an increased forging force at the end of the heating period (Lienert et al. 2012).

### 10.4.5.3 Heating Time

For a given application, heating time is determined during setup or from the previous experience. Extensive heating time limits the productivity of the line and increases materials wastage. Insufficient time leads to uneven heating, entrapment of oxides, and sometimes areas with poor bonding. Because of the geometrical reasons, near the center of the rotating bar heat generation would be less. Under such cases, thermal diffusion from the outer portion should occur to ensure uniform heating throughout the interface. Heating time will be significantly affected by heating (friction) pressure and tool rotation speed. Heating time will be reduced significantly as the heating pressure is increased.

### 10.4.6 Heat Generation in Friction Welding

Friction welding (FW) involves both heating and plastic deformation of the parent metal under extreme thermal and strain rate conditions (Zhang et al. 2006, 737). Moreover, the temperature rise is rapid and strain variation is also extreme. Both bring the complexity of high level in analyzing the temperature distribution. Also, extreme strain rate in the order of $10^3$/s brings adiabatic heating, which varies from point to point. It is now well accepted that modeling of heat flow phenomenon is a complex and accurate assumption of this at the faying interface is impossible (Grong and Grong 1997; Maalekian 2008, 1429). Also, the coefficient of friction "$\mu$" is continuously changing with time.

The frictional heat generation at the rubbing surface is more uniform in orbital friction welding compared to other types of friction welding. This uniformity is attributed to the uniform unidirectional relative velocity between the two components over the whole interfacial area. The higher integrity of the welds observed in orbital friction welding is attributed to this uniformity in heat generation. The heat generation term in the orbital friction welding is being modeled using four components: Coulomb friction which is supposed to be uniform, sliding–sticking friction, power data details, and inverse heat conduction approach. The modeled heat generation term is found to be quiet reasonable. For dissimilar welds used in critical applications, they are useful.

### 10.4.7 HAZ during Friction Welding

The HAZ consists of the following three specific zones (refer Figure 10.6):

A fully plasticized region (A): This is immediately next to the interface. Here the parts are expected to accommodate the plastic strain by dynamic recovery and recrystallization. A clear change in the microstructure will be noticed in this zone.

Partially deformed region (B): Here the degree of plastic deformation and temperature would be less compared to A. Because of the absence of recrystallization, increased dislocation density would be observed. Also, depending on the system, precipitates causing hardening would be dissolved. This is basically a thermo-mechanically affected zone.

Undeformed region (C): Here partial dissolution of the precipitates will occur. Since deformation does not take place, no change in the dislocation density would be seen. Since microstructural changes take place without any deformation, this is actual HAZ. Because of microstructural changes, mechanical properties also do change.

A sketch presented in Figure 10.6 indicates these three zones, in a situation where similar metallic materials are welded. Similar zones will form in a dissimilar metals combination. The scales of different zones will depend on the processing parameters.

## 10.5 Friction Welding of Pure Dissimilar Metals Combinations

Joining of dissimilar materials combination is relevant for those systems where a certain combination of properties is required. Also, it helps in saving costs and scarce resources. Conventional methods, like fusion welding, are not practical due to the formation of the brittle phases, the formation of low-temperature eutectic products, low melting point phases, metallurgical compatibility, thermal mismatch, etc. For such systems, solid-state welding in general and friction welding, in particular, is relevant.

Combination of metals which are pure help in understanding the mechanism of the joining which is easier to extend to complex systems (Meshram, Mohandas, and Reddy 2007, 330). There are many such applications. For example, Fe–Ni-based systems are relevant in aero-engine applications. Ti alloy–steel combination is relevant in cryogenic plumbing systems (Ji et al. 2016, 107; Cola et al. 1999, 962), aero and chemical industries (Ferranate and Pigoretti 2002, 2825). Copper and iron base alloys are used in the fabrication of cooling tubes for heat dissipation (Meshram, Mohandas, and Reddy 2007, 330) as well as power generation and transmission systems (Teker 2013, 303; Wang et al. 2013).

Fuji, Horiuchi, and Yamamoto (2005, 287) have investigated the strength behavior of the friction-welded pure Ti and pure Ni joints. They compared the strength of the as-welded and post-weld heat-treated samples. They noted that the reaction layer thickness increased with the increase in the friction time. The reaction layer thickness was more at the periphery compared to the central region. Aspects of dissimilar joint design are also investigated (Lin et al. 1999, 100s). They also noted that the material farthest from the center undergoes plastic flow. It is attributed to the increased tangential velocity at the periphery. This fills up the gap and prevents other parts from getting friction welded. An alternative suggested by the authors was to have a convex curvature (i.e., central axis will be bulging) for one of the faying surfaces.

### 10.5.1 Deformation Behavior

In general, the material having the lower strength within the combination selected, at that temperature, experiences more deformation and it results in higher level of flash formation. In Ti–Fe combination, Ti is stronger at lower temperature, but as the temperature increases, due to allotropic transformation involving hcp to beta change, strength drops and hence during welding the Ti deforms more compared to the iron. In the same way in Ni–Ti combination, Ti undergoes deformation. In Fe–Cu combination, Cu undergoes deformation. They are shown in the macrographs presented in Figure 10.7. In friction welding, heat is generated due to the rubbing of the faying surfaces. The amount of heat generated is proportional to the relative velocity. Relative velocity is minimum at the center and maximum at the periphery. However, at the periphery, heat is lost to the surrounding. Hence, the maximum temperature is reached at a point slightly below the surface. The difference in thermal diffusivities and cooling effects decide the overall temperature at a point. This affects the flash formation, whether one-sided, two-sided, equal, or non-equal.

### 10.5.2 Burn-Off and Reaction Layer Thickness

With increase in the burn-off time, the thickness of the reaction layer increases. The weld interface is generally straight. Figure 10.8a shows a uniform interface between pure Ni and pure Ti. This may not be true at the extremes, where considerable intermixing is possible (Figure 10.8b).

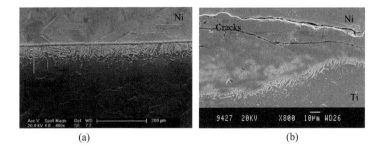

(a)                                             (b)

**FIGURE 10.8**
Micrographs indicate uniform reaction layer for the bulk of the section (a) and excessive thickness for the reaction layer leading to cracks at the periphery (b) Dissimilar combination : Pure Ni–Ti.

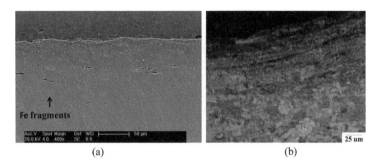

**FIGURE 10.9**
Mixing at the interface is indicated by material transport across the interface. Micrograph in (a) indicates the transfer of Fe into the copper side in a Fe–Cu weldment. Micrograph in (b) indicates the flow of soft metal Cu in Fe–Cu system.

More intermixing at the periphery of the weld region is attributed to increased relative velocity and higher temperature.

### 10.5.3 Heat-Affected Zone

In all systems, thermomechanical deformation or dynamic recrystallization is possible. During friction welding, the material near the interface undergoes frictional heating and plastic deformation. This favors dynamic recrystallization in low-strength materials. The width of this is more for low-strength materials (for Cu in Cu–Fe system). The extent of this will be different on two sides of the weldment. The magnitude depends on the thermophysical properties of the individual components.

### 10.5.4 Mixing at the Interface

Figure 10.9 presents the distribution of two materials in the form of alternate bands. This is attributed to mechanical transport of materials radially from the center. Conventional diffusion of elements to either side of the interface should not exceed a couple of micrometers. In practice, the observed scale of materials transport is more by an order. Again, the mechanism will be different depending on the type of systems (i.e., solid solution forming, phase separating, intermetallic forming, etc.)

## 10.6 Use of Friction Welding for Dissimilar Materials Joining

Most probably, the most important contribution of friction welding toward manufacturing is the possibility of joining a wide combination of dissimilar materials. In this count, friction welding outclasses all other welding processes. Also, the joining process being automatic, the productivity is high. Friction welding can join materials with great differences in mechanical, physical, and metallurgical properties.

### 10.6.1 Use of Friction Welding for Dissimilar Joints Involving Al Alloys

Friction welding has the ability to weld dissimilar materials and is found to be more suitable compared to any other solid-state methods (Yilbas et al. 1995, 76). It is more so, wherein one of the materials is aluminum-based alloy. It is observed that the oxide contamination in aluminum generally reduces the quality of the weld. This is also true with the other contaminants that do not permit close contact between the parts which is essential for ensuring complete bonding. Also, reducing the speed (rpm) and friction time reduces the quality of the joint. The effect of reduced rpm could partially be overcome by increasing friction time which in turn increases the burn-off length. Increase in rpm and burn-off length increases weld zone temperature, excessive flow of the material and hence loss of the material.

A wide combination of aluminum alloys could be friction welded. The literature clearly distinguishes two types of dissimilar combinations within aluminum alloys. In one case, the matrix is the same, whereas the difference arises due to a different grade. This combination is relatively simple to join, and it is possible to get good joints by friction welding. When the matrix itself is different, new problems arise mainly because of the formation of the intermetallics. The joint strength varies from case to case, which mainly depends on the formation of the intermetallics at the interfaces (Tokisue 2004, 861). But the strength is certainly better compared to fusion welding of same dissimilar materials combination, and rejection rate is also lower.

Joining of aluminum alloys to steel has been a challenging one, especially using fusion welding route. Much better results were reported using friction welding. Fukumoto et al. (1999, 1080), using continuous drive friction welding, joined 5052 aluminum alloy and 304 austenitic stainless steel. They observed that a longer friction time caused the formation of the excessive intermetallic layer, IML. Fukumoto et al. (2002, 219) studied the interface between friction-welded commercial pure aluminum and austenitic stainless steel. Their results showed the extensive plastic deformation of aluminum near the weld interface. The microstructure consisted of aluminum grains oriented from the center of the weldment. The aluminum near the joint interface was refined due to hot deformation. Dynamically recrystallized grains of size in the order of $1\,\mu$m were observed. These grains were larger in size at the periphery compared to the grains at the center of the weldment (Fukumoto et al. 2002, 219). The grain size increased with an increase in the friction time, which has resulted in a decrease in the hardness. During friction stage, aluminum got hot worked, and during the upset stage, it got dynamically recrystallized. After the upset stage, during cooling, dynamically recrystallized grains grow.

Sahin, Yilbas, and Al-Garni (1996, 89) have shown that in aluminum–steel dissimilar weldment, the temperature decay on the aluminum side is slower than that on the steel side. This resulted in the heat accumulation on the aluminum side, which results in the HAZ on the aluminum side. Also, the oxide layer on the aluminum side got broken and the fresh area was exposed during friction stage. Yilbas et al. (1995, 431) studied friction-welded aluminum–steel and aluminum–copper dissimilar joints. They found that tensile properties improved when the IML thickness is in the range of 0.2–1 µm, whereas after that increase in the IML thickness had resulted in the poor strength. The thickness of the IML was maximum at the mid-radius, and it was less on both sides. This is attributed to differential heating along the interface and convective effect on the rotating part.

In the case of aluminum–copper friction welding, an increase in the friction load and rotation speed increased IML thickness. Also, it was noted that a further increase in the IML thickness reduces the tensile strength.

Effect of friction load, the speed of rotation, and welding duration on the joint performance differ depending on the dissimilar materials combination. For example, for steel–aluminum joint, these parameters are insignificant as far as joint quality is concerned. But the effect was significant on the weld properties. Same parameters were found to be less significant on the aluminum–copper joint properties.

Imaz, Col, and Acet (2002, 421) noted that the interface study is very much essential to understand weld quality. For example, the tensile strength of the aluminum–interstitial free steel is appreciably less compared to that of the aluminum–stainless steel (Kimura et al. 2009a, 655) combination. Also, to be noted is that the strength drops with an increase in the friction time, true for both combinations.

An investigation by Yokoyama (2003, 308) indicates that the tensile properties of the friction-welded 6061 aluminum alloy and 304 stainless steel is affected by the processing parameters, especially the amount of forge pressure. Increasing forge pressure gave improved tensile properties. Light metals such as aluminum and magnesium alloys could be joined with different steels (Sahin 2009b, 487).

The extensive research indicates that optimum joint properties are obtained by using a set of process parameters. Since a large number of experiments are required, statistical analysis is an economical and reliable method for optimizing welding parameters. In the case of friction welding of austenitic stainless steel and aluminum parts, the joint strength increased and decreased as a function of increasing friction time. The presence of contaminants at the interface reduces the joint quality. Surface finish was not found to be an important parameter (Uday et al. 2010, 534).

Song et al. (2008, 14) investigated mechanical properties of the friction-welded aluminum–nodular iron combination. They observed that at the central region, no IMC layer forms, and at the periphery $Fe_xAl_y$-based IMCs, it forms. This is attributed to differential heat energy generation at different points along the interface. They noted that this has resulted in the dropping of the average strength value. Since IMCs ($Fe_xAl_y$ based) are brittle in nature, observed drop in strength is attributed to IMCs forming at the peripheral region.

### 10.6.2 Welding of Dissimilar Steels

Generally, welding of carbon and low-carbon steels is not problematic, and there are ways to do it successfully. But the same thing cannot be said about medium carbon and alloy steels. But they could be friction welded. The success depends on the strict control of parameters, because of the possibility of difficulties due to their hardening properties. The automobile sector is one major domain where friction-welded steel components are in use (Balasubramanian et al. 1999, 845; Gould 2011). For example, steel stampings are friction welded to the shoulder of steel rods. Automotive transmission input shaft assemblies, valves, and turbochargers are joined to steel using direct drive friction welding (Gould 2011). Similarly, nickel-based alloys are joined to steel compression wheels using friction welds, for their applications in the aerospace sector. Reddy and Rao (2009, 875) studied the microstructure and mechanical properties of the dissimilar joints made up of austenitic stainless steel (AISI 304), ferritic stainless steel (AISI 430), and duplex stainless steels. For comparison purposes, friction-welded samples were compared with the EB-welded samples. In EB-welded samples, central region consisted of fine grains and peripheral region consisted of coarse grains. On the duplex steel side, friction welds produced joints with equal proportions of the austenite and the ferrite. Diffusion of the elements across the interface was significantly less in friction weldments.

Satyanarayana and Mohandas (2003, 184) did a large number of experiments on steel to establish friction weld microstructure and process parameters. Coarse grain size, in general, produces maximum toughness and low hardness. Finer grain size improves the notch tensile strength. A high friction force, low forge force with a high burn-off combination maximizes the notch tensile strength. Alternatively, a combination of low friction force, low forge force, and a high burn-off leads to improved toughness.

Ozdemir (2005, 2504) friction welded AISI 304 L stainless steel with AISI 4340 as a function of rotation speed and studied microstructure and mechanical properties of the joint. Microstructure wise four zones were identified at and near the interface. They are parent zone, partially deformed zone (POZ), the deformed zone (DZ), the transformed and fully recrystallized and fully deformed zone (FPDZ). The width and geometry of these regions were a function of rotation speed. As far as mechanical properties are concerned, use of higher rotation speed and a shorter friction time increased the tensile strength of the friction-welded joint.

Satyanarayana, Reddy, and Mohandas (2005, 128) have friction-welded austenitic and ferritic stainless steels. Different thermal and physical properties of the materials (heat capacity, thermal conductivity, temperature, and hardness) have resulted in an asymmetric deformation. At a given temperature, austenitic stainless steel has lower thermal conductivity and greater hardness compared to ferritic stainless steel. For this reason, austenitic stainless steel does not undergo any deformation compared to ferritic stainless steel. Formation of flash is restricted to ferritic stainless steel only. A narrow band of deformation in the austenitic stainless steel suggests that it also undergoes deformation, but the extent of deformation is small and it does not lead to flash (collar) formation. Under low forging pressure combination, the degree of cold working is small and grain structure is coarse. Higher the forge pressure, the degree of working is more and it generates small grain size. The temperature of the peripheral region would be higher and hence the grain structure would be coarser at the periphery. The overall bond between the austenitic stainless steel and ferritic stainless steel is stronger than that between the friction-welded similar ferritic steels. Similarly, toughness is better than the ferritic stainless steel parent metals.

Satyanarayana, Reddy, and Mohandas (2005, 128) have studied the effect of variable surface roughness on the austenitic-ferritic stainless steel dissimilar welding. With increase in the surface roughness of the austenitic stainless steel faying surfaces, the extent of deformation features on both sides increased, though austenitic stainless steel did not take part in the formation of the flash. When the surface roughness is high, the banded features consisting of alternate layers of austenite and ferrite layers were observed near the interface. The hardness of the interface and the adjacent austenite stainless steel increased with increasing roughness and it is attributed to increased deformation.

### 10.6.3  Friction Welding of Steels with Other Metals

In friction welding, heat generated due to the controlled rubbing of the contact surfaces is harnessed to create a joint. Based on the literature, we can say that a large number of process parameters do affect the quality of the joint. Since the literature on the friction welding of dissimilar welding is rare, use of conventional techniques for optimization is difficult.

Experiments by Sahin et al. (1998, 127) on friction welded copper–steel joints indicated a similarity on the temperature profile and diffusion profile, along with the radial direction along the interface. Diffusion profile affects the concentration profile and hence joint quality. The temperature reaches its maximum value at far from the center but not at the free surface. This variation is expected due to the difference in heat generation as well as due to the difference in heat dissipation. Also, the width of the HAZ is affected by the temperature variation both in the radial and axial directions. The HAZ is found to be wider in the higher-diffusivity region compared to the lower-diffusivity side. The temperature developed is higher in the outer region. This results in relatively large melting and HAZ developing in this region. Copper side (in copper-steel) exhibited small grains, and it is attributed to the cold working.

Dey et al. (2009, 704) have friction welded 304L stainless steel to Ti. It may be noted that this is an important combination of materials in the design of systems for processing of spent fuels in nuclear industry.

Kimura et al. (2009b, 704) have considered the effect of friction welding condition on the strength of the brass to the low-carbon steel joint. For analysis, they had varied friction welding parameters, like friction pressure, friction time, and forge pressure. They noted that at about 30 MPa friction pressure, the brass pieces were transferred to the low-carbon steel side at mid-center (half-radius) location. The quantity of the transfer and the area over which it transferred increased as a function of friction time.

### 10.6.4 Copper and Copper Alloys

Copper and its alloys have high thermal conductivity and due to that heat rapidly diffuses into the base metal. Because of the difficulty in concentrating the energy at and near the interface, generally, weldability of materials with high thermal conductivity is difficult especially using fusion weld processes. Such problems as in copper alloys are easily overcome using friction welding.

---

## 10.7 Friction Welding of Ceramics

Ceramic materials as a group have been highly promising due to the excellent strength, high wear, and corrosion resistance, over a wide range of operating parameters. But difficulties in processing, like machining, have put a brake on their applicability. This makes joining of ceramics to ceramics as well as ceramics to metals, a domain for development. Hence, friction welding is a highly promising technique (Kanayama et al. 1985, 95).

Figure 10.10 sketches the configuration during ceramics to metal joining. Here the ceramic material is held stationary, while the soft metal is rapidly rotated. Rubbing a metal against a hard stationary piece enables the plastic deformation of the metal. Point to be noted is that the pressure should not exceed the strength of the ceramic part. Otherwise, the ceramic part will fracture. People have successfully joined aluminum to alumina and magnesium to yttria-stabilized zirconium oxide ceramics. If joining is done under a protective environment the metal oxidation could be avoided (Nascimento, Martinelli, and Buschinelli 2003, 178). Similarly to avoid large thermal gradient adjacent to faying surfaces (Nascimento, Martinelli, and Buschinelli 2003, 178), aluminum inserts and pipe were used. They protect the ceramic component from thermal shock and stress concentration.

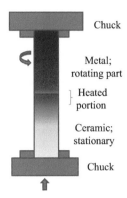

**FIGURE 10.10**
Configuration during friction welding of ceramic to the metallic part.

Joining of alumina to mild steel by friction welding is another combination having some interest. This combination has all the inherent limitations of metal–ceramic joints listed already. They also have an additional limitation in the form of the possibility of a reaction between them at the interface (leading to iron aluminum spinel). An important feature of the metal–ceramic dissimilar joint is associated with the difference in hardness, melting, and coefficient of thermal expansion. The large inherent difference in coefficient of thermal expansion (CTE) between ceramic–metal combination makes this joint to have large induced thermal stresses at the interface leading to joint failure. To overcome the CTE mismatch, a gradient material made up of metal/alloy or metal–ceramic component is inserted. Results indicate that in such joints, joint strength depends on the aluminum interlayer thickness and mechanical locking at the interface on both sides of the interlayer. They also noted that an increase in the friction time leads to an increase in the bending strength (Mohamad, Luay, and Zainal 2007, 1016; Ahmad Fauzi et al. 2010, 670).

In another study, Ahmad Fauzi et al. (2010, 670) studied the interface of friction-welded alumina and 6061 aluminum alloy, from an angle of quality of the bonding. They investigated the effect of process parameters on the degree of deformation. It was noted that the effect of rotational speed is high on the 6061 aluminum alloy than on the alumina part. It was noted that the weld interface had three zones—first, unaffected base metal zone, followed by deformation zone and finally fully recrystallized and fully deformed zone (FPDZ). When the rotational speed is increased, the thickness of the FPDZ at the interface increased as a result of more mass thrown out from the welding interface.

In a ceramic material, the heat absorption is slow due to low thermal conductivity, in the contact area, a large temperature gradient is created generating large thermal stresses. During cooling to room temperature, residual stresses are generated due to the disparity between the coefficients of thermal expansion of two materials. This generates thermal-based stresses which are disastrous for the integration of the joint.

In a metal–metal friction joint, both elements are plastically deformed near the weld interface (depending on parameters), and sometimes they may even form new phases and compounds (Zimmerman 2006, 462). Contrast to that ceramic materials remains elastic whereas aluminum deforms plastically in the weld zone and their stress–strain plot depends on the temperature.

## 10.8 Friction Welding of Polymers

For joining of polymers, the principles applicable to metal–metal combination are applicable. Here also, one part is fixed, while the other part is rotated against it with a controlled velocity. Frictional heat is generated at the interface causes the polymer to melt and a weld is formed during cooling. Major welding parameters include rotational speed, weld time, friction pressure, forge pressure, and burn-off the length. The process produces a weld of high quality, simplicity, and reproducible. Since plastics are poor conductors of heat, frictional heat generated at the interface is accumulated at the interface. Depending on the combination, one part may get heated up rapidly and due to pressure, a bond is formed without softening of the interior of the part. The investigation showed that the HAZ has three parts: the plasticized region, partly plasticized region, and the undeformed region (Singh et al. 2016, 77; Kah et al. 2014, 152).

## 10.9 Summary

This chapter reviews the challenges and difficulties involved in friction welding of dissimilar materials, covering metals and alloys, ceramics, and polymers. Although fundamentals are the same, for different classes of materials, the domain of parameters for getting reliable performance is different. For getting better results, a correlation between processing parameters, microstructure development, and joint needs to be done. Extensive research and data are required for each class of engineering materials. Also, investigations on microstructural development at the interface and its correlation to mechanical properties will help. Use of numerical tools like finite-elemental analysis, experimental techniques, and optimization tools will accelerate the research in this domain.

## References

Ageorges, C., and Ye, L. 2001. "Resistance welding of thermosetting composites/thermoplastic composites." *Composites, Part A-Applied Science and Manufacturing,* 32: 1603–1621.

Ahmad Fauzi, M.N., Uday, M.B., Zuhailwati, H., and Ismail, A.B. 2010. "Microstructure and mechanical properties of alumina 6061 aluminum alloy joined by friction welding." *Materials & Design,* 31(2): 670–676.

Alves, E.P., An, C.Y., Pierino, F., Eduardo, N., and do Dantos, F. 2012. "Experimental determination of temperature during rotary friction welding of dissimilar materials." *Frontiers in Aerospace Engineering,* 1: 20–26.

Balasubramanian, V., Stotler, T., Li, Y., Crompton, J., Soboyejo, A., Katsube, N., and Soboyejo, W. 1999. "A new friction law for the modeling of continuous drive friction welding: Application to 1045 steel welds." *Materials and Manufacturing Processes,* 14(6): 845–860.

Balle, F., Wagner, G., and Eifler, D. 2009. "Ultrasonic metal welding of aluminium sheets to carbon fiber reinforced thermoplastic composites." *Advanced Engineering Materials,* 11: 35–39.

Cola, M.J., Lyons, M.B., Theter, D.F., and Gentzlinger, R.C. 1999. "Dissimilar metal joints for the APT superconducting cavity's cryogenic plumbing system." *Proceedings of the 1999 Particle Conference,* New York, 962–964.

D'Alvise, L., Massoni, E., and Walloe, S.J. 2002. "Finite element modeling of the inertia friction welding process between dissimilar materials." *Journal of Materials Processing Technology,* 125–126(9): 387–391.

Daus, F., Li, H.Y., Baxter, G., Bray, S., and Bowen, P. 2007. "Mechanical and microstructural assessments of RR1000 and IN 718 inertial welds effects of welding parameters." *Materials Science and Technology,* 23: 1424–1432.

Dey, H.C., Ashafq, M., Bhaduri, A.K., and Rao, K.P. 2009. "Joining of titanium to 304L stainless steel by friction welding." *Journal of Materials Processing Technology,* 209(18–19): 5862–5870.

El-Hardek, M. 2014. "Numerical simulation of inertial friction welding process of dissimilar materials." *Metallurgical and Materials Transactions B,* 45B: 2346–2356.

Fabritsiev, S.A., Pokrovsky, A.S., Nakamichi, M., and Kawamura, H. 1998. "Irradiation resistance of DS copper/stainless steel joint fabricated by friction welding methods." *Journal of Nuclear Materials,* 258–263: 2030–2035.

Ferranate, M., and Pigoretti, E.V. 2002. "Diffusion bonding of Ti-6Al-4V to AISI 316L stainless steels: Mechanical resistance and interface microstructure." *Journal of Materials Science,* 37: 2825–2833.

Filho, S.T.A., and dos Satos, S.T. 2009. "Joining of polymers and polymer-metal hybrid structures: Recent developments and trends." *Polymer Engineering & Science,* 49(8): 1461–1476.

Fuji, A., Horiuchi, Y., and Yamamoto, K. 2005. "Friction welding of pure titanium and nickel." *Science and Technology of Welding and Joining,* 10(3): 287–294.

Fukumoto, S., Tsubakino, H., Okita, K., Aritoshi, M., and Tomita, T. 1999. "Friction welding process of 5052 aluminum alloy to 304 stainless steel." *Materials Science and Technology,* 15(9): 1080–1086.

Fukumoto, S., Tsubakino, H., Tomita, T., and Okita, K. 2002. "Dynamic recrystallization phenomena of commercial purity aluminum during friction welding." *Materials Science and Technology*, 18(2): 219–225.

Gould, J.E. November 10, 2011. "Welding of dissimilar materials combinations for automotive applications." https://energy.gov/sites/prod/files/2013/11f4/multimaterial_joining_gould_ewi.pdf. Accessed Feb 15, 2018.

Grong, O., and Grong, D. 1997. *"Metallurgical Modeling of Welding."* London, Institute of Materials.

Imaz, M.Y., Col, M., and Acet, M. 2002. "Interface properties of aluminium-steel friction welded components." *Materials Characterization*, 49(5): 421–429.

Jamaludin, S.B., Keat, Y.C., and Ahmad, Z.A. 2004. "The effect of varying process parameters on the microhardness and microstructure of Cu-Steel and Al-Al$_2$O$_3$ friction joints." *Jurnal Teknologi*, 41(A): 85–95.

James, J.A., and Sudhish, R. 2016. "Study on effect of interlayer in friction welding for dissimilar steels: SS304 and AISI1040." *Procedia Technology*, 25: 1191–1198.

Ji, Y., Wu, S., and Zhao, D. 2016. "Microstructure and mechanical properties of friction welding joints with dissimilar titanium alloys." *Metals*, 6: 107–117.

Kah, P., Suoranta, R., Martikainen, J., and Magnus, C. 2014. "Techniques for joining dissimilar materials." *Reviews on Advanced Materials Science*, 36: 152–164.

Kanayama, K., Tasaki, Y., Machida, M., Kume S., and Aoki, S. 1985. "Joining of ceramics by friction welding." *Transactions of the Japan Welding Society*, 16(1): 95–96.

Kim, W.S., Yun, I.H., Lee, J.J., and Jung, H.T. 2010. "Evaluation of mechanical interlock effect on adhesion strength of polymer-metal interfaces using micropatterned surface topography." *International Journal of Adhesion and Adhesives*, 30(6): 408–417.

Kimura, M., Ishil, H., Kusaka, M., Kaizu, K., and Fuji, A. 2009a. "Joining phenomena and joint strength of friction welded joint between aluminium-magnesium alloy (AA5052) and low carbon steel." *Science and Technology of Welding and Joining*, 14: 655–661.

Kimura, M., Kasuya, K., Kausaka, M., Kaizu, K., and Fuji, A. 2009b. "Effect of friction welding condition on joining phenomena and joint strength of friction welded joint between brass and low carbon steel." *Science and Technology of Welding and Joining*, 14(5): 704–712.

Kimura, M., Seo, K., Kusaka, M., and Fuji, A. 2003. "Observation of joining phenomena in friction stage and improving friction welding method." *JSME International Journal (Series A)*, 46A(3): 384–390.

Kurt, A., Uygur, I., and Paylasan, U. 2011. "Effect of friction welding parameters on mechanical and microstructural properties of dissimilar AISI 1010-ASTM B22 joints." *Welding Research*, 90: 102s–108s.

Li, H.Y., Huang, Z.W., Bray, S., Baxter, G., and Bowen, P. 2007. "High temperature fatigue of friction welded joints in dissimilar nickel based superalloys." *Materials Science and Technology*, 23: 1408–1418.

Lienert, T., Siewert, T., Babu, S., and Acoff, V. (eds). 2012. *"ASM Handbook Volume 6A: Welding Fundamentals and Processes."* Materials Park, OH: ASM International.

Lin, C.B., Mu, C.K., Wu, W.W., and Hung, E.H. 1999. "The effect of joint design and volume fraction on friction welding properties of A360/SiCp composites." *Welding Journal*, 78: 100s–108s.

Maalekian, M. 2007. "Friction welding critical assessment of the literature." *Science and Technology of Welding and Joining*, 12(8): 738–759.

Maalekian, M. 2008. "Thermal modeling of friction welding." *ISIJ International*, 48(10): 1429–1433.

Maalekian, M., Kozeschnik, E., Brantner, H.P., and Cerjak, H. 2008. "Comparative analysis of heat generation in friction welding of steel bars." *Acta Materialia*, 56(12): 2843–2855.

Meshram, S.D., Mohandas, T., and Reddy, G.M. 2007. "Friction welding of dissimilar materials." *Journal of Materials Process Technology*, 184: 330–337.

Mohamad, Z.N., Luay, B.H., and Zainal, A.A. 2007. "Alumina–mild steel friction welded at lower rotational speed." *Journal of Materials Process Technology*, 10: 1016–1021.

Mortensen, K.S., Jensen, C.G., Conrad, L.C., and Losee, F. 2001. "Mechanical properties and microstructures of inertia welded 416 stainless steel." *Weld Journal Research Supplement*, 268–273.

Mousavi, S., and Kelishami, A.R. 2008. "Experimental and numerical analysis of the friction welding process for the 4340 steel and mild steel combinations." *Welding Journal*, 87(7): 178s–186s.

Nagatsuka, K., Yoshida, S., Tsuchiya, A., and Nakata, K. 2015. "Direct joining of carbon fiber-reinforced plastic to an alumina alloy using friction lap joining." *Composites Part B*, 73: 82–88.

Nascimento, R.M., Martinelli, A.E., and Buschinelli, A.J.A. 2003. "Review article: Recent advances in metal-ceramic brazing." *Ceramics*, 49, 178–198.

Nguyen, T.C., and Weckman, D.C. 2006. "A thermal and microstructure evolution model of direct drive motion welding of plain carbon steel." *Metallurgical and Materials Transactions B*, 37B(2): 275–292.

Nicholas, E.D. 2003. "Friction processing technologies." *Weld World*, 47(11–12): 2–9.

O'Brien, R.L. (ed). 1991. *"Welding Handbook v2, Welding Processes."* Miami, FL, American Welding Society.

Ozdemir, N. 2005 "Investigation of the mechanical properties of friction welded joints between AISI 304L and AISI 4340 steel as a function of rotational speed." *Materials Letters*, 59(19–20): 2504–2509.

Reddy, G.M., and Rao, K.S. 2009. "Microstructure and mechanical properties of similar and dissimilar stainless steel electron beam and friction welds." *International Journal of Advanced Manufacturing Technology*, 45(9): 875–888.

Reddy, M.G., Rao, S.A., and Mohandas, T. 2008. "Role of electroplated interlayer in continuous drive friction welding of AA 6061 to AISI304 dissimilar metals." *Science and Technology of Welding and Joining*, 13: 619–628.

Sahin, A.Z., Yibes, B.S., Ahmed, M., and Nickel, J. 1998. "Analysis of the friction welding process in relation to the welding of copper steel bars." *Journal of Materials Processing Technology*, 82(1): 127–136.

Sahin, A.Z., Yilbas, B.S., and Al Garni, A.Z. 1996. "Friction welding of Al-Al, Al-steel and steel-steel samples." *Journal of Materials Engineering and Performance*, 5(1): 89–99.

Sahin, M. 2007. "Evaluation of the joint interface properties of austenitic stainless steels (AISI 304) joined by friction welding." *Materials and Design*, 28(7): 2244–2250.

Sahin, M. 2009a. "Joining of stainless steels and copper materials with friction welding." *Industrial Lubrication and Tribology*, 61(6): 319–324.

Sahin, M. 2009b. "Joining of stainless steel and aluminium materials by friction welding." *International Journal of Advanced Manufacturing Technology*, 41(5): 487–497.

Sakiyama, T., Naito, Y., Miyazaki, Y., Nose, T., Murayama, G., Saita, K., and Oikawa, H. 2013. "Dissimilar metal joining technologies for steel sheet and aluminum alloy sheet in autobody." Nippon Steel Technical Report, May, No 103.

Satyanarayana, V.V., and Mohandas, T. 2003. "Continuous drive friction welding studies on AISI 430 ferritic stainless steel." *Science and Technology of Welding and Joining*, 8(3): 184–193.

Satyanarayana, V.V., Reddy, G.M., and Mohandas, T. 2005. "Dissimilar metal friction welding of austenitic-ferritic steels." *Journal of Materials Processing Technology*, 160(2), 128–137.

Singh, R., Kumar, R., Feo, L., and Fraternali, F. 2016. "Friction welding of dissimilar plastic/polymer materials with metal powder reinforcements for engineering applications." *Composites Part B*, 101: 77–86.

Song, Y.L., Liu, Y.H., Zhu, X.Y., Lu, S.R., and Zhang, Y.B. 2008. Strength distribution at rotary friction welded aluminum to nodular cast iron." *Transactions of Nonferrous Metals Society of China*, 18(1): 14–18.

Teker, T. 2013. "Evaluation of the metallurgical and mechanical properties of friction welded joints of dissimilar metal combination AISI 2205/Cu." *International Journal of Advanced Manufacturing Technology*, 66: 303–310.

Tenaglia, R. 2012. "Dissimilar materials joining." https://ewi.org/dissimilar_materials_joining. Accessed Oct 10, 2017.

Tokisue, K.K.H. 2004. "Dissimilar friction welding of aluminium alloys to other materials." *Welding International*, 18(11): 861–867.

Uday, M.B., Ahmad Fauzi, M.N., Zuhailawati, H., Ismail, A.B. 2010. "Advances in friction welding process—A review." *Science and Technology of Welding and Joining*, 15(7): 534–558.

Wang, Y., Luo, J., Wang, X., and Xu, X. 2013. "Interfacial characterization of T3 copper/35CrMnSi steel dissimilar metal joints by inertia radial friction welding." *International Journal of Advanced Manufacturing Technology*, doi: 10.1007/s00170-013-4936-7.

Winiezenko, R., and Kaczorowski, M. 2012. "Friction welding of ductile cast iron using interlayers." *Materials and Design*, 34: 444–451.

Yilbas, B.S., Sahin, A.Z., Coban, A., and Aleem, B.J.A. 1995. "Investigation into the properties of friction welded aluminium bars." *Journal of Materials Processing Technology*, 54(1–4): 76–81.

Yilbas, B.S., Sahin, A.Z., Kahraman, N., and Al-Garni, A.Z. 1995. "Friction welding of steel-Al and Al-Cu materials." *Journal of Materials Processing Technology*, 49(3–4): 431–443.

Yokoyama, T. 2003. "Impact tensile properties of friction welded butt joints between 6061 aluminum alloy and type 304 stainless steel." *JSME International Journal (Series A)*, 46A: 308–315.

Zhang, Q.Z., Zhang, L.W., Liu, W.W., Zhang, X.G., Zhu, W.H., and Qu, S. 2006. "3D rigid viscoplastic FE modeling of continuous drive friction welding process." *Science and Technology of Welding and Joining*, 11(6): 737–743.

Zimmerman, J. 2006. "Thermo-mechanical phenomena in friction welding $Al_2O_3$ ceramic to aluminum." *Welding International*, 20(6): 462–465.

# 11

## Insights into the Flux-Assisted TIG Welding Processes

Jaykumar J. Vora

*Pandit Deendayal Petroleum University*

### CONTENTS

11.1 Introduction.................................................................................................237
11.2 Activated TIG Welding Technique.........................................................238
    11.2.1 A-TIG Welding with Different Fluxes on Various Steels.........240
11.3 Flux-Bounded TIG Welding Technique..................................................245
11.4 Flux-Zoned TIG Welding...........................................................................246
11.5 Insights into Depth-Enhancing Mechanisms......................................247
11.6 Summary .......................................................................................................254
References.................................................................................................................255

## 11.1 Introduction

The manufacturing activity as an industry is always the key parameter for the development of any nation. The large-scale lucrative businesses of fabrication of vessels for oil and gas, automotive and its ancillaries, aerospace sectors, etc. are always critical and challenging task to perform. A considerable amount of research, care, and inputs are required for the successful delivery of the component. Thus, welding is one of the most challenging tasks that needs special attention in the manufacturing sector. The successful welding now calls for optimum quality, reduced cost, and faster completion. Number of conventional welding processes exist are in operation since long; however, while selecting the process, a careful consideration is needed on the final application of the component. Several applications require very high-quality joints such as pipes and components for oil refining such as pressure vessels and boilers (Dhandha and Badheka 2015). These components operate at high pressure and temperatures, and hence any predicament on its safe working will be fatal. In general, tungsten inert gas (TIG) welding process is a widely accepted welding process wherein high-quality joints are intended. However, the major disadvantage of the TIG welding process is its relatively low penetration capability, particularly in single-pass autogenous (without filler metal mode) operations. The maximum weld penetration with the autogenous TIG welding is restricted to 3.0–3.5 mm in a single pass, making the welding of components of 6 mm or higher impossible to weld in a single pass (Vasantharaja and Vasudevan 2012, Naik et al. 2017, Vaishnani et al. 2017). The manufacturers hence resort to either double-sided welding which is not possible every time due to accessibility issue from the other side, design constraints, and challenging component handling (for reversing) in over dimensional vessels or carry out the welding from one side by adding filler metal into the groove which involves implications of cost, time, and energy. Thus, a need for devising a variant of TIG welding which can solve the problem was sensed.

A partial solution to the problem was the possibility of using chemical powder termed as flux, along with TIG welding which can increase the penetration capability of the process and make it usable. The idea behind using flux with TIG welding was imbibed from the characteristic sensitivity of the TIG process to the base material chemical composition in form of the cast-to-cast variation. Heiple and Burgardt (1993) reported that in the same metal composition, different batches of the same steel exhibited different penetration profiles. The chemical composition of the batches of this steel was the same; however, a minor change in chemical composition of trace elements was present. It was concluded that TIG welding is sensitive to the cast-to-cast variations, and hence the chemical composition (especially of trace elements) plays a pivotal role in determining the weld penetration during the TIG welding. This quality of the TIG welding was used to increase the penetration by applying the flux on the top surface of the plate to be welded so that fundamentally a change in chemical composition is induced. Since the inception of this technology, various flux-assisted TIG welding techniques have surfaced, primarily different in the technique of applying the flux. There are three major types of the flux-assisted TIG welding wherein a considerable amount of study has been conducted, namely activated TIG (A-TIG) welding, flux-bounded TIG (FB-TIG) welding, and flux-zoned TIG (FZ-TIG) welding.

In this chapter, an attempt has been made to introduce the concept and fundamentals of A-TIG, FB-TIG, and FZ-TIG welding techniques; summarize the application of the techniques on different steels; and evaluate their performance in increasing the weld penetration. This penetration increase by flux application is due to the different depth-enhancing mechanism prevalent during the welding specific to the flux and base metal used. The chapter also provides an insight into this mechanism.

## 11.2 Activated TIG Welding Technique

Activated TIG welding process was first introduced in the 1960s by Paton Welding Institute, Ukraine. A simple process of applying a thin layer of flux

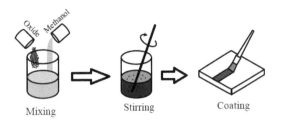

**FIGURE 11.1**
Flux application methodology.

over the surface of the component to be welded before welding showed an outstanding increment in the weld penetration as compared to conventional autogenous TIG welding at same parameters (Vora and Badheka 2017). The flux was a chemical powder comprising primarily of metal oxides, fluorides, halides, etc., which was applied as a thin paste layer prior to welding on the metal surface as shown in Figure 11.1. In the flux-assisted TIG techniques, the flux available in the powdered form is not possible to apply evenly on the weld surface. For this purpose, the powder is converted into the paste form by mixing it with acetone as shown in Figure 11.1a and applying it on the intended area (generally throughout the length of the plate) as shown in Figure 11.1b. Acetone has a tendency to vaporize quickly leaving the evenly distributed oxide flux on the surface. Several other techniques have been proposed by the different authors; however, the mentioned technique is one of the widely preferred and experimented methods.

After the coat is dry, autogenous TIG welding is carried out on the surface of the plate with a fine layer of the flux in between arc and base metal as shown in Figure 11.2. During welding, this fine layer of flux is melted and vaporized at arc temperatures and penetration capability is increased by up to 300% when compared to conventional TIG process (Vora and Badheka 2016a).

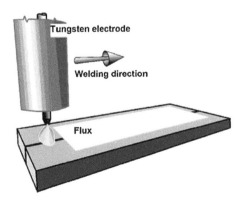

**FIGURE 11.2**
A-TIG welding technique (Dhandha and Badheka 2015). (With kind permission from Elsevier.)

The increase in penetration was largely attributed to the different depth-enhancing mechanisms prevalent during the A-TIG welding process which has also been summarized in the subsequent sections. This capability enabled steel of around 6 mm to be welded with a single pass in square butt configuration without the addition of filler metal. A-TIG welding is thus able to produce enhanced weld penetration in single-pass welding technique and henceforth simplifying joining procedures.

### 11.2.1 A-TIG Welding with Different Fluxes on Various Steels

Since its inception, various studies have been reported for the use of A-TIG welding technology. As the technique is correlated with the chemical composition of the flux as well as the chemistry of the steel on which A-TIG is carried out, hence varied results were obtained specific to the different types of steel, fluxes, and responsible depth-enhancing mechanism, as summarized in Table 11.1. The fluxes that can be used in A-TIG welding can either be primarily comprised of oxides, chlorides, bromides, halides, or sulfides and known to us as activating fluxes (Vora et al. 2017). Also, the fluxes can be used either as a single component and can be mixed in a different compositions. Researchers have also reported the use of a mixture of three different fluxes in A-TIG welding. In addition to this, a variety of steels such as stainless steels (SS), alloy steels, titanium, and magnesium have been attempted for A-TIG welding. A consolidated summary of the outputs of the work done with different types of fluxes on various steels by the researchers has been summarized herewith.

Out of all the published research articles on A-TIG welding, the majority of the articles are pivoted to the joining of stainless steel and its alloys. A number of fluxes were used for A-TIG welding of stainless steel and its alloys, and it was proposed that the effects of these fluxes were different from each other while keeping other parameters same for all. A simple reason is that arc welding of SS and its alloy always possesses a challenge due to the phenomenon of sensitization and hot cracking. Any welding process involving a reduced number of passes, elimination of filler metal, and reduced heat input shall be favored for its joining. In addition to this, it is a well-known fact that SS and its alloy pose a problem of distortion and residual stresses on welding. As a result, A-TIG welding of SS was attempted largely, and various studies reported that a reduction of distortion, as well as residual stresses, and adequate corrosion resistance were achieved for the SS weldments. Shyu et al. (2008) investigated the effect of oxide fluxes on weld morphology, angular distortion, and hot cracking susceptibility obtained with autogenous TIG welding, applied on 5-mm-thick SS 304 austenitic stainless steel plates. Single-component oxide powders such as $Al_2O_3$, $Cr_2O_3$, $TiO_2$, $SiO_2$, and CaO were used for the study. Experimental results indicated that the increase in the penetration was significant with the use of $Cr_2O_3$, $TiO_2$, and $SiO_2$. It was also reported that A-TIG welding can increase the weld-depth to bead-width

**TABLE 11.1**

Review of Work Carried Out on A-TIG Welding

| Author and Year | Base Metal | Thickness (mm) | Fluxes | Responsible Mechanism |
|---|---|---|---|---|
| Tseng and Hsu (2011) | SS 316L | 6 | $MnO_2$, $TiO_2$, $MoO_3$, $SiO_2$, and $Al_2O_3$ | Reversed Marangoni effect and arc constriction |
| Huang (2009) | SS 304 | 5 | $MnO_2$ and ZnO | Not specified |
| Yang et al. (2003) | 0Cr18Ni9 | 6 | $SiO_2$ and $TiO_2$ | Reversed Marangoni effect and arc constriction |
| Modenesi, Apolinário, and Pereira (2000) | SS 304 | 5–8 | Mixed | Reversed Marangoni |
| Tseng and Wang (2014) | SS 316L | 5 | $FeS/FeS_2$ | Reversed Marangoni effect and arc constriction |
| Li et al. (2007) | 1Cr18Ni9Ti | Varies | $SiO_2$ and $TiO_2$ | Reversed Marangoni effect and arc constriction |
| Lu et al. (2002) | SS 304 | 10 | $Cu_2O$, NiO, $Cr_2O_3$, $SiO_2$, and $TiO_2$ | Revered Marangoni |
| Zhang, Pan, and Katayama (2011) | SS 304 | 12 | $Al_2O_3$, $Fe_2O_3$, $SiO_2$, $Cr_2O_3$, $TiO_2$, $MnO_2$, and $B_2O_3$ | Reversed Marangoni effect, arc constriction, and Lorentz force |
| Tseng (2013) | SS 316L | 6 | $SiO_2$, $TiO_2$, $Cr_2O_3$, $MoO_3$, $NiF_2$, and $MoS_2$ | Reversed Marangoni effect and arc constriction |
| Huang et al. (2005) | SS 304 | 5 | $MnO_2$ and ZnO | Arc constriction and anode spot |
| Shyu et al. (2008) | SS 304 | 5 | $Al_2O_3$, CaO, $TiO_2$, $Cr_2O_3$, and $SiO_2$ | Arc constriction |
| Tseng and Chen (2012) | SS 316L | 6 | $SiO_2$ and $TiO_2$ | Reversed Marangoni effect and arc constriction |
| Chern, Tseng, and Tsai (2011) | 2205 Duplex stainless steel | 6 | $TiO_2$, $MnO_2$, $SiO_2$, $MoO_3$, and $Cr_2O_3$ | Arc constriction and anode spot |
| Ramkumar, Kumar, et al. (2015) | Inconel 718 | 5 | $SiO_2$ and $TiO_2$ | Reversed Marangoni effect and arc constriction |
| Ramkumar, Varma, et al. (2015) | SS 904L | 5 | $SiO_2$ and $TiO_2$ | Reversed Marangoni effect and arc constriction |

*(Continued)*

**TABLE 11.1 (*Continued*)**

Review of Work Carried Out on A-TIG Welding

| Author and Year | Base Metal | Thickness (mm) | Fluxes | Responsible Mechanism |
|---|---|---|---|---|
| Tathgir, Bhattacharya, and Bera (2015) | AISI 1020<br>SS 304<br>SS 316<br>2205 Duplex stainless steel | Varies | $TiO_2$ | Reversed Marangoni effect and arc constriction |
| Modenesi et al. (2015) | SS 304 | 5 | $Cr_2O_3$, $KCLO_4$, and $Al_2O_3$ | Dilution effect |
| Liu et al. (2015) | SS 304 | 8 | $SiO_2$, $Cr_2O_3$, $Al_2O_3$, CuO, NiO, $TiO_2$ and $MnO_2$ | Reversed Marangoni effect and arc constriction |
| Vora and Badheka (2015) | RAFMS | 6 | $Al_2O_3$, $Co_3O_4$, CuO, HgO, $MoO_3$ and NiO | Reversed Marangoni effect and arc constriction |
| Vora and Badheka (2016b) | RAFMS | 6 | CaO, $CrO_3$, $Fe_2O_3$, $MnO_2$, $TiO_2$, and ZnO | Reversed Marangoni effect and arc constriction |
| Cai et al. (2016) | BS700MC | 5 | $B_2O_3$ and $Cr_2O_3$ | Reversed Marangoni effect and arc constriction |
| Li et al. (2012) | AZ31 | 6 | $CaF_2$ and $TiO_2$ | Reversed Marangoni |
| Ramkumar, Dev, et al. (2015) | Dissimilar (Inconel 718 and AISI 416) | 5 | $TiO_2$ | Arc constriction |
| Nayee and Badheka (2014) | Dissimilar (carbon steel and SS) | 6 | $TiO_2$, ZnO, and $MnO_2$ | Reversed Marangoni effect and arc constriction |
| Kumar and Sathiya (2015) | Incoloy 800H | 4 | $SiO_2$ and ZnO | Reversed Marangoni |

ratio, and tends to reduce the angular distortion of the weldment. An increase in the retained delta-ferrite content of SS welds was reported which consequently reduced the hot cracking susceptibility. Dong et al. (2004) investigated the effect of eight types of fluxes; out of those, five were a single component ($TiO_2$, $Cr_2O_3$, $SiO_2$, $ZrO_2$, and $AlF_3$), and three were multicomponent on penetration depth and welding arc profiles in conventional TIG welding of SS 304 stainless steel. The weld penetration depth was dramatically increased with the use of single-component oxide fluxes such as $Cr_2O_3$, $SiO_2$, and $TiO_2$. Furthermore, in a similar study by Huang et al. (2005), the effect of specific oxide fluxes on the surface appearance, weld morphology, retained ferrite content, hot cracking susceptibility, and angular distortion with the A-TIG welding process applied to the welding of 5-mm-thick SS 304 plates. However, the uniqueness of the study was the use of multicomponent fluxes wherein a mixture of 80% $MnO_2$–20% ZnO fluxes imparted full penetration and also a satisfactory surface appearance to A-TIG welds. TIG welding with $MnO_2$ and/or ZnO increased the measured ferrite number in welds and reduced hot cracking susceptibility in as-welded structures. It was also found that TIG flux welding significantly reduced the angular distortion of SS weldment. A comparison of A-TIG welding with multipass TIG welding was presented by Vasantharaja, Vasudevan, and Palanichamy (2015). They prepared weld joints of 10-mm-thick 316LN stainless steel made by multipass TIG and A-TIG welding processes. The residual stress distribution and distortion values were measured and compared. It was reported that the weld joints made by TIG and A-TIG welding processes exhibited different microstructures, peak tensile residual stress, and angular distortion values. The A-TIG weld joint exhibited lower peak residual stress and distortion values due to lower weld metal volume caused by the absence of filler metal addition and straight sided edge preparation, single-pass welding, and more intense heat source. In one of the studies, Tseng (2013) reported A-TIG welding of SS316L steel with an indigenously developed activated flux by the National Pingtung University of Science and Technology (NPUST). The flux was primarily oxide-based flux powder in addition to fluorides and sulfides ($SiO_2$, $TiO_2$, $Cr_2O_3$, $MoO_3$, $MoS_2$, and $NiF_2$), and its effect on the surface appearance, geometric shape, angular distortion, and ferrite content of the SS welds was investigated. For the investigated currents of 125–225 A, the maximum penetration of stainless steel activated TIG weld was obtained when the coating density was between 0.92 and 1.86 $mg/cm^2$. An important aspect of the study was the selection of carrier solvent. In major studies, acetone is used for converting the flux into a paste; however, the authors reported the use of methanol as a carrier solvent. It was observed that methanol imparted good spreadability to the flux paste which enhanced the results. The depth of finger-like profile in the conventional TIG weld increased in conjunction with the current because of the induced strong arc pressure. The arc pressure also raised the penetration capability of activated TIG welds at high currents. The results indicated that higher current levels have lower ferrite content of austenitic 316L stainless steel weld metal than lower current levels. In another study, five kinds of oxide fluxes, $MnO_2$, $TiO_2$, $MoO_3$, $SiO_2$, and $Al_2O_3$, were used to investigate the effect of A-TIG process on weld morphology, angular distortion, delta-ferrite content, and hardness of Type 316L stainless steels by Tseng and Hsu (2011). An autogenous TIG welding was applied to 6-mm-thick stainless steel plates through a thin layer of flux to produce a bead-on-plate welded joint. The experimental results indicated that the $SiO_2$ flux facilitated root pass joint penetration, but $Al_2O_3$ flux led to the deterioration in the weld depth and bead width compared with conventional TIG process. It was concluded that A-TIG welding can increase the joint penetration and weld depth-to-width ratio, thereby reducing angular distortion of the weldment. Paulo et al. (2000) studied six types of fluxes including oxides and fluorides, namely, $Al_2O_3$, $AlF_3$, $CaF_2$, $Cr_2O_3$, $Fe_2O_3$, $SiO_2$, and $TiO_2$ at the same time on A-TIG welding of SS 304. It was detected that the high penetration occurred just after the transition from TIG to A-TIG welding. Some fluxes ($Cr_2O_3$, $SiO_2$, and $TiO_2$) gave exceptionally good penetration.

Another predicament that surfaced was the effect of the amount of flux particles on enhancing the weld depth penetration (Tseng 2013; Ahmadi and Ebrahimi 2015; Modenesi et al. 2015; Bachmann et al. 2013). The flux effect on TIG weld shape variations was investigated by the aid of the heat transfer and fluid flow model combined with experimental trials A-TIG welding of Nimonic 263 alloy, and TiO, $TiO_2$, and $Ti_2O_3$ as the fluxes. It was proposed that by controlling the quantity of the active elements, i.e., the quantity of flux, different kinds of the weld shapes were obtained. The experimental results showed that the increase of active flux on the weld bead tends to increase the penetration of the weld pool at first and then decreases steeply. Thus, a definite value of the flux termed as "critical quantity" of flux was proposed which governs the penetration in A-TIG welding. Above the value, weld penetration was not increased, and below the value, enhanced penetration was achieved (Xu et al. 2007). A similar observation was reported during A-TIG welding of SS 304 stainless steel by single-component oxide fluxes such as $Cu_2O$, $NiO$, $Cr_2O_3$, $SiO_2$, and $TiO_2$ wherein weld depth-to-width ratio initially increased with the increase in the quantity of flux and later on decreased (Lu et al. 2002).

Several encouraging studies have also been carried out on A-TIG welding of alloy steels as well as dissimilar combinations. Liu et al. (2007) studied the effect of five single-oxide fluxes on the depth-to-width ratio in AZ31B magnesium alloy. The aim of the study was to analyze the role of oxygen in the deeper penetration. The oxygen content in the weld seam was measured and the arc images during the TIG welding process were captured. An increase in penetration was reported for all five fluxes using the alternating current TIG (AC-TIG) welding process. When compared the process with no-flux TIG welding, both the welds showed the marginal difference in the oxygen content. However, welds that have the best penetration had a relatively higher oxygen content among those produced without flux. Vora and Badheka

(2015, 2016b) carried out the studies on A-TIG welding of reduced activation ferritic martensitic steels (RAFM) wherein 12 different oxide fluxes were attempted. It was inferred from the studies that few fluxes gave deeper penetration than the normal TIG weld weldments and few fluxes gave drastically increase in penetration (greater than even plate thickness). The authors suggested the combination of more than one depth-enhancing mechanism present for the process. Thus, A-TIG welding has been experimented rigorously on different steels.

## 11.3 Flux-Bounded TIG Welding Technique

Followed by the success of the A-TIG welding, several efforts were directed toward devising more improved techniques to encounter the limitations of A-TIG. Flux-bounded TIG welding (FB-TIG) technique is one such variant wherein a small gap in the middle of the plate (intended area) is retained and the flux is coated on either side of this gap as shown in Figure 11.3 The gap is widely referred to as the flux gap and is an important parameter for the success of this technique.

The technique was first proposed by Sire and Marya (2001) wherein it was identified that the selective flux application has the capability of shaping the welding arc as required. The need for the development of this technique was its capability to weld magnesium or other lower melting point alloys to avoid excessive vaporization loss of alloy elements due to high heat input in conventional arc welding processes. Though fundamentally A-TIG and FB-TIG welding are both similar, however in FB-TIG, the flux layer does not fully cover weld joint but is applied on the borders of the abutting surfaces leaving behind a narrow length of exposed metal on which the autogenous TIG weld arc is moved as shown in Figure 11.3 Since its realization, several

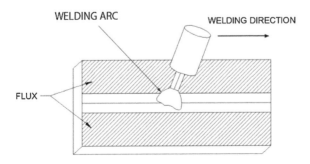

**FIGURE 11.3**
FB-TIG technology (Jayakrishnan and Chakravarthy 2017). (With kind permission from Elsevier.)

studies have been carried out on different steels, flux types, and thicknesses which are summarized below.

The first investigation performed by Sire and Marya (2001) was on 6-mm-thick aluminum plates using FBTIG by maintaining a flux gap of 4 mm. The authors achieved deeper welds as compared to conventional TIG welding welds at the same welding parameters. As reported by the authors, the increase in power density was diverted toward the narrow channel created using the FB-TIG welding flux application technique. Further studies carried out by Sire, Rückert, and Marya (2002) and Sire and Marya (2002) proved their case wherein the weld penetration and depth-to-width ratio for FB-TIG were almost double than that obtained for normal TIG weld. Zhao et al. (2011) showed that most of the typical single-component fluxes showed an enhanced penetration for various possible flux gap values with silica giving the maximum penetration. The phenomenon proposed by them was the reversed Marangoni. Rückert et al. (2013) and Rückert, Huneau, and Marya (2007) proposed that the flux gap played a vital role in dictating the weld penetration. Another phenomenon "Melting Point Depression" was proposed by Jayakrishnan, Chakravarthy, and Rijas (2017) wherein finer particle size of the activating flux yielded deeper penetration. This resulted in a higher concentration of the surface active elements in the weld pool for the same welding conditions. A comprehensive study by Santhana Babu et al. (2014) established that the tensile properties of the FB-TIG welded joints were more than that of conventional welds with about 10% increase. Even the stress corrosion cracking resistance of FBTIG-welded specimen was comparable to that of a conventional TIG-welded specimen.

Even though all the depth-enhancing mechanisms prevalent in A-TIG welding are proposed to be prevalent with FB-TIG also, the mechanisms inactive in A-TIG are present in the FB-TIG and which needs further studies and consideration. One such major contributing mechanism for FB-TIG welding is the insulation effect, wherein the arc caves in, resulting in a higher energy density at the arc root which has been explained in a subsequent section.

## 11.4 Flux-Zoned TIG Welding

The latest variant to surface in the flux-assisted TIG welding technique was the flux-zoned TIG (FZ-TIG). It can be devised as a combination of A-TIG as well as FB-TIG welding techniques. FZ-TIG welding was proposed by Huang, Fan, and Shao (2012) by altering the activating flux coating method, based upon the arc constriction theory. As shown in Figure 11.4 in FZ-TIG welding, a thin activating flux layer with low melting temperature, boiling temperature, and current resistivity is applied in the central part, whereas the thin flux layer with high melting temperature, boiling temperature, and

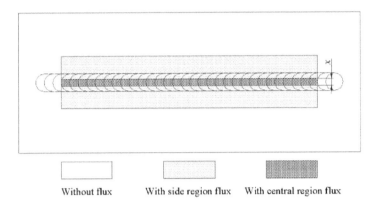

**FIGURE 11.4**
FZ-TIG technology (Huang, Fan, and Shao 2012). (With kind permission from Taylor & Francis.)

current resistivity is painted on the central and side regions of the weld surface. This is followed by the carrying out conventional autogenous TIG welding. Given the novelty of the technique, the literature and studies reported in the open domain are extremely scarce.

The idea seed for the development of this technology was imbibed from the understanding of the different depth-enhancing mechanisms and the need to arc weld aluminum alloys which are otherwise quite difficult. This is due to the fact that the thermophysical properties of aluminum have low melting and boiling temperature, and high thermal conductivity leading to the low temperature gradient of the weld pool. Due to this low surface tension, aluminum alloys become easy to oxidize and by generating a thin layer of oxide film. This makes the change in the sign of the weld pool surface tension temperature coefficient from positive to negative through difficult which is normally carried out by introducing active elements to lower down the surface tension of molten metal. This makes the strong inward flow phenomenon difficult to prevail and hence limit the weld penetration. However, the same can be encountered by applying the layer of two different fluxes with the difference in their thermophysical properties. Using the FZ-TIG welding process, the difficult problem of simultaneously improving the weld surface appearance and increasing the weld penetration can be solved as shown in Figure 11.5 wherein the weld surface appearance is comparable with the TIG welding and the weld penetration is drastically increased.

## 11.5 Insights into Depth-Enhancing Mechanisms

Certain instances were reported wherein autogenous TIG welding with activated flux resulted in an intense enhancement in the depth of penetration,

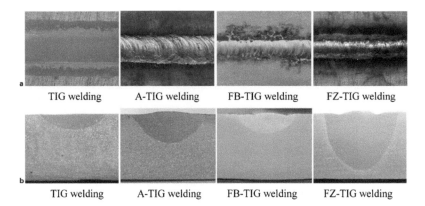

TIG welding        A-TIG welding        FB-TIG welding        FZ-TIG welding

TIG welding        A-TIG welding        FB-TIG welding        FZ-TIG welding

**FIGURE 11.5**
Comparison of flux-assisted TIG welding techniques (Huang, Fan, and Shao 2012). (With kind permission from Taylor & Francis.)

reduced weld width, and subsequent increase in weld depth-to-width ratio as compared to the conventional TIG process with the same welding conditions. In line with this, studies pivoted to knowing the reason for the enhancement of this penetration was in focus. Since then, several researchers identified various mechanisms and forces acting during the welding which might have increased the penetration. The proposed depth-enhancing mechanisms are as follows (Heiple and Burgardt 1993; Gas Tungsten Arc Welding 1993; Takeuchi, Takagi, and Shinoda 1992; Marya 1996; Chase and Savage 1971):

- Arc constriction due to negative ions or insulating flux layer
- Reversed Marangoni effect
- Buoyancy or gravity forces
- Aerodynamic drag forces
- Electromagnetic or Lorentz force

Majority of the studies in the open literature are carried out on the A-TIG welding; however, it is worthwhile to note that the depth-enhancing mechanism is almost same in all; however, those absent in the A-TIG welding are activated in FB-TIG and FZ-TIG.

Reversed Marangoni effect and arc constriction effect are the two mechanisms widely proposed as the depth-enhancing mechanism proposed by researchers. During initial studies, the enhancement of penetration by A-TIG welding was suggested to be due to the arc constriction phenomenon. However, certain studies didn't report the arc constriction mechanism and yet deeper penetration was achieved during the A-TIG welding (Tanaka et al. 2000; Modenesi, Apolinário, and Pereira 2000). Thus, this contrasting results motivated the researchers on carrying out the in-depth studies to identify the probable depth-enhancing mechanisms, as this would lead to increased usability of this newly emerging welding process (Li et al. 2014; Huang 2009, 2010; Niagaj 2003; Modenesi, Apolinário, and Pereira 2000; Howse and Lucas 2000). The subsequent results by researchers' indicated that the weld bead shape and size is the result of the fluid flow of the molten metal during welding. Any changes in the fluid flow during the welding will have a resulting effect on the weld bead. Thereafter, an effect termed as reversed Marangoni effect which encompasses changes in fluid flow was identified as the dominant depth-enhancing mechanism (Lu, Fujii, and Nogi 2004; Lu et al. 2003; Marya and Edwards 2002; Mills and Keene 1990; Saidov et al. 2000; Vora and Badheka 2016).

The arc constriction mechanism was widely proposed as the depth-enhancing mechanism in A-TIG welding of various steels with various fluxes. The arc constriction mechanism was reported to be caused due to two different aspects. The electron attachment at the edge of the arc forms negative ions which cause arc constriction and an increase in current density at the anode center. The decomposing flux either comprising oxides, halides or fluorides will further increase the magnitude of electrons and a bright anode spot (Lowke, Tanaka, and Ushio 2005).

Another reason for the arc constriction was reported due to the insulating surface of flux. In A-TIG welding, the surface of the workpiece is covered with a thin layer of flux, which is usually a metal oxide. This layer acts as an insulation barrier to the arc current. Temperatures at the center of the weld pool are sufficient enough to melt the flux so that the electric current can penetrate the flux to the weld puddle and further to the workpiece. The arc diameter at the weld pool surface will be reduced by the insulating effect of the flux for the outer regions of the arc. Fundamentally, the arc constriction takes place in A-TIG welding with the help of fluxes and the relevant mechanisms as compared to mechanical constriction done in Plasma Arc Welding (PAW). For a given current, the current density at the center of the weld pool increases leading to increased magnetic pinch forces and pressure in the weld pool, resulting in strong convective flow downwards in the weld pool and an increased weld depth. Lowke, Tanaka, and Ushio (2005) reported that in the high-temperature region which is the center of the weld pool, the insulating flux is evaporated to allow arc current to enter the weld pool. The A-TIG welding carried out on the coated surface reduces the diameter of the plasma arc column. The anode root area increases the current density and hence the heat density increases, which produces narrower and deeper weld than conventional TIG welding. A clear distinct change in the arc images can be observed.

The reversed Marangoni effect was another widely proposed depth-enhancing mechanism in A-TIG welding of various steels with various fluxes. The normal tendency of the surface tension is that it decreases with increase in temperature; in other words, $\partial\sigma/\partial T < 0$ for a pure metal and many alloys. In the weld pool for such materials, surface tension is higher in the relatively

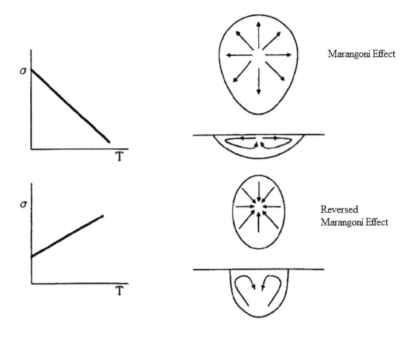

**FIGURE 11.6**
Marangoni vs. reversed Marangoni effect (Berthier, Paillard, and Christien 2009). (With kind permission from Taylor & Francis.)

cooler parts of the pool edge than it is in the pool center under the arc. Hence, the fluid flows from the pool center to the edge, and the heat flux easily transfers to the edge and forms a wide and shallow weld shape. Dissolved flux in the molten weld pool can change surface tension properties. In particular, a change from a negative to a positive gradient of the surface tension with temperature can change the surface convective flow in the weld pool from being radially outward to radially inward as shown in Figure 11.6. Radially inward flow along the surface of the weld pool leads to downward flow at the center of the weld pool, and this downward flow leads to heat being carried downwards, thus resulting in a deeper weld (Lowke, Tanaka, and Ushio 2005). These were the two widely proposed depth-enhancing mechanism, and several researchers reported their existence with several pieces of evidence which are summarized herewith.

Cai et al. (2016) carried out the A-TIG welding with two different fluxes $B_2O_3$ and $Cr_2O_3$ on BS700MC steel and reported an in-depth study of the responsible mechanism for deeper penetration. The authors observed that penetration was enhanced with both fluxes $B_2O_3$ and $Cr_2O_3$; however, the extent of increase was substantial with flux $B_2O_3$. After an in-depth analysis, it was concluded that simultaneous effect of two different depth-enhancing mechanisms (reversed Marangoni and arc constriction) and the enhanced penetration with $Cr_2O_3$ was only due to the reversed Marangoni effect. The presence of the reversed Marangoni was confirmed initially with the typical low-width high-depth weld bead achieved with the use of both fluxes individually. To ensure their claim authors added a tracing element tungsten (W) within the flux and its presence was confirmed after the welding. It was observed that the W was more at the bottom of the bead than at the surface in A-TIG welding contrary to conventional TIG welding wherein the W was more on the edges. Hence, it was concluded that due to a change in fluid flow the deeper penetration was achieved.

Furthermore, the arc images observed by the authors were a clear indication of the presence of arc constriction mechanism present with flux $B_2O_3$. However, the authors also proved that the arc voltages with the A-TIG welding were increased with the use of flux $B_2O_3$ only. Nearly similar arc voltage was observed for conventional TIG welding and A-TIG welding with flux $Cr_2O_3$. Thus, it can be concluded that flux $B_2O_3$ triggered two different depth-enhancing mechanisms and hence gave a substantial increase in penetration.

In a similar study on A-TIG welding of SS 304, use of nine different oxide fluxes was reported wherein the change in penetration was associated with the change in arc voltage. Arc voltage is expected to remain the same during TIG and A-TIG welding as no changes in arc length were reported. However, a variation of up to 3V was recorded with the use of fluxes, and hence enhanced penetration was observed (Modenesi, Apolinário, and Pereira 2000).

It was also suggested that a small amount of impurities or elements in the composition of steel can result in large changes in arc characteristics and the surface properties of the weld pool (Chase and Savage 1971; Mills and Keene 1990; Takeuchi, Takagi, and Shinoda 1992). The presence of surface active elements like sulfur and oxygen in the steel causes these alterations in surface tension. Tseng and Hsu (2011) carried out the A-TIG welding of SS 316L stainless steel using five kinds of single-component oxide fluxes, $MnO_2$, $TiO_2$, $MoO_3$, $SiO_2$, and $Al_2O_3$. It was inferred from the studies that $SiO_2$ flux facilitated root pass joint penetration, but $Al_2O_3$ flux led to the deterioration in the weld depth and bead width compared with conventional TIG process. However, it was worth noting that all the trials were taken at the same welding conditions. Similarly, the weld depth-to-width ratio of the A-TIG weld bead was increased leading to the reduction in the angular distortion as compared to autogenous TIG weld bead. The reason for the enhanced penetration with A-TIG welding was proposed as centripetal Marangoni convection (reversed Marangoni) which is related to the surface tension fluid flow of the molten metal. The surface active elements present in the flux such as oxygen reverses the usual tendency of the fluid to get drifted toward the edges and draws the molten metal at the center enhancing the weld depth and reducing the weld bead width (Modenesi, Apolinário, and Pereira 2000; Tanaka et al. 2000; Dong et al. 2004; Berthier et al. 2012). A deep and narrow fingerlike penetration was obtained during A-TIG welding with $SiO_2$ flux.

Leconte et al. (2006) focused on oxide fluxes of A-TIG welding. It appeared from the studies that oxides have two different effects: one on the fusion zone chemistry and the other on the electric arc behavior. First, oxygen reverses the Marangoni convection movements that become centripetal, thus contributing to increase 'D' penetration. On the other hand, oxides may cause an increase in the energy flux density transferred by the arc to the metal.

Oxide powders $Al_2O_3$, $Cr_2O_3$, $TiO_2$, $SiO_2$, and CaO were used for the study. Experimental results indicated that the increase in the penetration is significant with the use of $Cr_2O_3$, $TiO_2$, and $SiO_2$. It was also reported that welding can increase the weld-depth to bead-width ratio, and tends to reduce the angular distortion of the weldment. It was also found that A-TIG welding can increase the retained delta-ferrite content of stainless steel 304 welds and, in consequence, the hot cracking susceptibility was reduced. Physically constricting the plasma column and reducing the anode spot were reported as a possible mechanism. In a similar study, Yushchenko, Kovalenko, and Kovalenko (2006) performed a comparative analysis of TIG and A-TIG over stainless steel SS 304. The study revealed the amount of oxygen, surface tension values, temperature gradients, and the direction of fluid flow as the major influencing parameters.

Tseng and Wang (2014), in a slightly different approach, evaluated the potential of using TIG welding assisted by a mixture of ionic fluxes $FeS/FeF_2$ to improve joint penetration and weld quality in joining SS 316L plates. The differences in surface appearance, geometric shape, and out-of-plane deformation of the weldments made by using conventional and A-TIG welding processes were compared. The results show that the surface discontinuities on the welded surface which are at times reported to be inferior by authors were decreased by increasing the concentration of $FeF_2$ in $FeS/FeF_2$ mixtures. This undercutting was eliminated because when $FeS/FeF_2$ mixture was used, the molten metal completely filled the melted out areas of the base metal. The results also indicated that A-TIG welding using $75\%FeS/25\%FeF_2$ produced a greater increase in the weld depth-to-width ratio, resulting in a significant reduction of weld-induced deformation in stainless steel weldments. Thus, apart from several single-component combinations of flux, a mixture of flux was also developed for effective A-TIG welding of SS.

As conclusively explained above, apart from reversed Marangoni effect and arc constriction mechanism, certain other phenomenon is reported by researchers which might be the cause for enhanced penetration. Tseng and Shiu (2015) reported that the fluid flow dominates energy transfer within the molten pool, and thus determines the shape and size of the weld. During the arc welding, the liquid metal flow within the molten pool is driven primarily by the gas shear stress (SS) exerted by the plasma jet (JP), the Marangoni force (FM), the Lorentz force (FL), and the buoyancy force (FB) as shown in Figure 11.7. The actions of the driving forces were based on the net energy balance of the electric arc and molten pool and the difference in the distribution of static gas pressure between the electrode and the workpiece produces

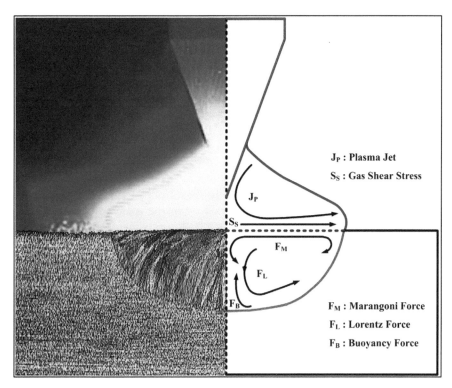

**FIGURE 11.7**
Different forces in A-TIG welding (Tseng and Shiu 2015). (With kind permission from Elsevier.)

a JP toward the workpiece during TIG welding. When the JP impinged on the molten pool, it induces the SS acting on the surface of the molten pool. Thus, the SS also influences the shape and size of the resulting weld. In addition, the FB was insignificant during the autogenous TIG welding of steel; it was therefore ignored by the authors.

Additionally, several mechanisms affecting the penetration mechanism during A-TIG welding, namely, are electromagnetic or Lorentz forces, buoyancy forces, and aerodynamic drag forces (Takeuchi, Takagi, and Shinoda 1992; Mills and Keene 1990; Tseng and Shiu 2015; Chase and Savage 1971). Electromagnetic forces are primarily caused by the interaction of the induced magnetic field and that of the current passes through a conductor. The welding current induces a magnetic field around the conductor, thus resulting in a Lorentz force which acts toward the weld pool center in the direction of current flow and hence enhancing the penetration. Additionally, buoyancy forces as reported by several researchers are caused by the density difference of the molten metal in the weld pool. The molten metal density decreases with increasing the temperature and hence, as a matter of fact, the fluid tends to flow from high density to low density. During A-TIG welding, it acts toward the surface of the weld pool and hence may be responsible for deeper penetration. Another force reported by authors responsible for deeper penetration was the aerodynamic drag force. Aerodynamic drag forces produced by the action of the arc plasma flow over the surface of the weld pool which induces an outward flow along the surface of the weld pool.

Thus, it can be proposed that the fluid flow in the weld pool is extremely complicated because reversed Marangoni, electromagnetic forces, buoyancy forces, and aerodynamic forces can all affect the flow. The condition even worsens given the different chemistry of the steels as well as the different fluxes and their concentrations. However, it can be concluded assertively that the reversed Marangoni force was predominant and has a major effect on the weld bead geometry (Oreper and Szekely 1984; Vora 2017).

In TIG welding, the energy for the melting of the metal is obtained from the kinetic energy of arc constituents such as electrons and ions. The flux primarily oxides and inorganic compounds are fundamentally poor conductors of electricity. Thus, during the FB-TIG welding, the resistance offered by the flux gap (as compared to flux layer) is less and charged carriers tend to follow least resistance path (Zhao et al. 2011). This channels the entire arc energy to the flux gap, and hence the arc is confined to that specific area. This phenomenon is known as insulation effect. Jayakrishnan and Chakravarthy (2017) also established from their comprehensive review studies that one of the prime mechanisms behind the increased weld penetration was the insulation effect. This effect is dominant only in FB-TIG, and this can be a probable reason for enhanced penetration capability of FB-TIG welding.

## 11.6 Summary

From the summarized facts, it can be observed that flux-assisted TIG welding processes increase the penetration capability of autogenous TIG welding process which renders the welding of mid-thickness components (6–10 mm) easy and reliable to weld. In order to weld these components in the absence of the flux-assisted TIG welding techniques, the weld joints need edge preparation and adding of filler wire—consumable incorporating time, cost, and monetary obligations. The benefits of this technique are even more pronounced for the welding of aluminum and magnesium alloys. The arc welding of these alloys is extremely difficult due to their oxide-forming tendency owing to high heat input. With the flux-assisted TIG welding processes, deeper penetration can be obtained and also the loss of alloying elements (due to a lower melting point) can also be avoided. From the above survey, it can also be concluded that flux is one of the most important parameters in deciding the success of the processes and the prevalent depth-enhancing mechanisms also need a careful consideration.

Thus, these techniques have a strong potential to be inducted in the mainstream fabrication industries; however, the full potential of welding technology is yet to be harnessed due to several challenges. The primary challenge is that different fluxes (oxides, halides, etc.) behave uniquely with different steels. No universal flux or flux combination (single component or a mixture) is available for a particular class of steels. This is due to the fact that these techniques are a function of the chemical constituent of steel and the flux applied. Thus, development of these technologies for a new steel demands extensive research on the flux and flux combination to give enhanced penetration. This is also partly due to the fact that limited knowledge is found in the open literature regarding the depth-enhancing mechanisms prevalent during the welding. Apart from these, there are reported literatures wherein inferior mechanical and metallurgical properties are achieved through the use of A-TIG welding technology. Thus, A-TIG welding can be deemed an exciting field of research in coming times. However, the major concern in using these technologies is its two-step process and the manual mode of application of the flux.

## References

Ahmadi, E, and AR Ebrahimi. 2015. "Welding of 316L austenitic stainless steel with activated tungsten inert gas process." *Journal of Materials Engineering and Performance* 24 (2):1065–1071.

Bachmann, M, V Avilov, A Gumenyuk, and M Rethmeier. 2013. "About the influence of a steady magnetic field on weld pool dynamics in partial penetration high power laser beam welding of thick aluminium parts." *International Journal of Heat and Mass Transfer* 60:309–321.

Berthier, A, P Paillard, and F Christien. 2009. "Structural and chemical evolution of super duplex stainless steel on activated tungsten inert gas welding process." *Science and Technology of Welding and Joining* 14 (8):681–690.

Berthier, A, P Paillard, M Carin, S Pellerin, and F Valensi. 2012. "TIG and A-TIG welding experimental investigations and comparison with simulation: Part 2-arc constriction and arc temperature." *Science and Technology of Welding and Joining* 17 (8):616–621.

Cai, Y, Z Luo, Z Huang, and Y Zeng. 2016. "Influence of oxides on microstructures and mechanical properties of high-strength steel weld joint." *High Temperature Materials and Processes* 35 (10):1047–1053.

Chase, TF, and WF Savage. 1971. *The Effect of Anode Composition on Tungsten Arc Characteristics.* Ann Arbor, MI: University Microfilms.

Chern, TS, KH Tseng, and HL Tsai. 2011. "Study of the characteristics of duplex stainless steel activated tungsten inert gas welds." *Materials & Design* 32 (1):255–263.

Dhandha, KH, and VJ Badheka. 2015. "Effect of activating fluxes on weld bead morphology of P91 steel bead-on-plate welds by flux assisted tungsten inert gas welding process." *Journal of Manufacturing Processes* 17:48–57.

Dong, C, Y Zhu, G Chai, Hui Zhang, and S Katayama. 2004. "Preliminary study on the mechanism of arc welding with the activating flux." *Aeronautical Manufacturing Technology, Supplement* 6:271–278.

Gas Tungsten Arc Welding. 1993. "Fluid Flow Phenomena during Welding."

Heiple, CR, and P Burgardt. 1993. "Fluid flow phenomena during welding." *ASM International, ASM Handbook* 6:19–24.

Howse, DS, and W Lucas. 2000. "Investigation into arc constriction by active fluxes for tungsten inert gas welding." *Science and Technology of Welding and Joining* 5 (3):189–193.

Huang, HY. 2009. "Effects of shielding gas composition and activating flux on GTAW weldments." *Materials & Design* 30 (7):2404–2409.

Huang, HY. 2010. "Research on the activating flux gas tungsten arc welding and plasma arc welding for stainless steel." *Metals and Materials International* 16 (5):819–825.

Huang, HY, SW Shyu, KH Tseng, and CP Chou. 2005. "Evaluation of TIG flux welding on the characteristics of stainless steel." *Science and Technology of Welding & Joining* 10 (5):566–573.

Huang, Y, D Fan, and F Shao. 2012. "Alternative current flux zoned tungsten inert gas welding process for aluminium alloys." *Science and Technology of Welding and Joining* 17 (2):122–127.

Jayakrishnan, S, and P Chakravarthy. 2017. "Flux bounded tungsten inert gas welding for enhanced weld performance—A review." *Journal of Manufacturing Processes* 28:116–130.

Jayakrishnan, S, P Chakravarthy, and AM Rijas. 2017. "Effect of flux gap and particle size on the depth of penetration in FBTIG welding of aluminium." *Transactions of the Indian Institute of Metals* 70 (5):1329–1335.

Kumar, SA, and P Sathiya. 2015. "Experimental investigation of the A-TIG welding process of Incoloy 800H." *Materials and Manufacturing Processes* 30 (9):1154–1159.

Leconte, S, P Paillard, P Chapelle, G Henrion, and J Saindrenan. 2006. "Effect of oxide fluxes on activation mechanisms of tungsten inert gas process." *Science and Technology of Welding & Joining* 11 (4):389–397.

Li, D, S Lu, D Li, and Y Li. 2014. "Principles giving high penetration under the double shielded TIG process." *Journal of Materials Science & Technology* 30 (2):172–178.

Li, QM, XH Wang, ZD Zou, and WU Jun. 2007. "Effect of activating flux on arc shape and arc voltage in tungsten inert gas welding." *Transactions of Nonferrous Metals Society of China* 17 (3):486–490.

Li, SZ, J Shen, ZM Cao, LZ Wang, and N Xu. 2012. "Effects of mix activated fluxes coating on microstructures and mechanical properties of tungsten inert gas welded AZ31 magnesium alloy joints." *Science and Technology of Welding and Joining* 17 (6):467–475.

Liu, GH, MH Liu, YY Yi, YP Zhang, ZY Luo, and L Xu. 2015. "Activated flux tungsten inert gas welding of 8 mm-thick AISI 304 austenitic stainless steel." *Journal of Central South University* 22 (3):800–805.

Liu, LM, ZD Zhang, G Song, and L Wang. 2007. "Mechanism and microstructure of oxide fluxes for gas tungsten arc welding of magnesium alloy." *Metallurgical and Materials Transactions A* 38 (3):649–658.

Lowke, JJ, M Tanaka, and M Ushio. 2005. "Mechanisms giving increased weld depth due to a flux." *Journal of Physics D: Applied Physics* 38 (18):3438.

Lu, S, H Fujii, and K Nogi. 2004. "Marangoni convection in weld pool in $CO_2$-Ar-shielded gas thermal arc welding." *Metallurgical and Materials Transactions A* 35 (9):2861–2867.

Lu, S, H Fujii, H Sugiyama, M Tanaka, and K Nogi. 2002. "Weld penetration and Marangoni convection with oxide fluxes in GTA welding." *Materials Transactions* 43 (11):2926–2931.

Lu, S, H Fujii, H Sugiyama, M Tanaka, and K Nogi. 2003. "Effects of oxygen additions to argon shielding gas on GTA weld shape." *ISIJ International* 43 (10):1590–1595.

Marya, SK. 1996. "Effect of minor chemistry elements on GTA weld fusion zone characteristics of a commercial grade titanium." *Scripta Materialia* 34 (11):1741–1745.

Marya, M, and GR Edwards. 2002. "Chloride contributions in flux-assisted GTA welding of magnesium alloys." *WELDING JOURNAL-NEW YORK-* 81 (12):291–298.

Mills, KC, and BJ Keene. 1990. "Factors affecting variable weld penetration." *International Materials Reviews* 35 (1):185–216.

Modenesi, PJ, ER Apolinário, and IM Pereira. 2000. "TIG welding with single-component fluxes." *Journal of Materials Processing Technology* 99 (1):260–265.

Modenesi, PJ, P Colen Neto, E Roberto Apolinário, and K Batista Dias. 2015. "Effect of flux density and the presence of additives in ATIG welding of austenitic stainless steel." *Welding International* 29 (6):425–432.

Naik, A, K Darshan, S Sagar, JJ Vora, V Patel, S Das, and R Patel. 2017. "Development of activated TIG welding technology for low alloy steels: A step towards sustainable manufacturing. In book Technology Drivers: Engine for Growth." *Proceedings of the 6th Nirma University International Conference on Engineering (Nuicone 2017)*, November 23–25, 2017, Ahmedabad, India.

Nayee, SG, and VJ Badheka. 2014. "Effect of oxide-based fluxes on mechanical and metallurgical properties of dissimilar activating flux assisted-tungsten inert gas welds." *Journal of Manufacturing Processes* 16 (1):137–143.

Niagaj, J. 2003. "The use of activating fluxes for the welding of high-alloy steels by A-TIG method." *Welding International* 17 (4):257–261.

Oreper, GM, and J Szekely. 1984. "Heat-and fluid-flow phenomena in weld pools." *Journal of Fluid Mechanics* 147:53–79.

Ramkumar, KD, S Dev, V Saxena, A Choudhary, N Arivazhagan, and S Narayanan. 2015. "Effect of flux addition on the microstructure and tensile strength of dissimilar weldments involving Inconel 718 and AISI 416." *Materials & Design* 87:663–674.

Ramkumar, KD, BM Kumar, MG Krishnan, S Dev, AJ Bhalodi, N Arivazhagan, and S Narayanan. 2015. "Studies on the weldability, microstructure and mechanical properties of activated flux TIG weldments of Inconel 718." *Materials Science and Engineering: A* 639:234–244.

Ramkumar, KD, JLN Varma, G Chaitanya, A Choudhary, N Arivazhagan, and S Narayanan. 2015. "Effect of autogeneous GTA welding with and without flux addition on the microstructure and mechanical properties of AISI 904L joints." *Materials Science and Engineering: A* 636: 1–9.

Rückert, G, B Huneau, and S Marya. 2007. "Optimizing the design of silica coating for productivity gains during the TIG welding of 304L stainless steel." *Materials & Design* 28 (9):2387–2393.

Rückert, G, N Perry, S Sire, and S Marya. 2013. "Enhanced weld penetrations in GTA welding with activating fluxes case studies: Plain carbon & stainless steels, titanium and aluminum." *THERMEC 2013* 202.

Saidov, R, H Mourton, R Le Gall, and G Saindrenan. 2000. "A-TIG welding of UR 52N+ superduplex stainless steel." *Welding International* 14 (8):633–639.

Santhana Babu, AV, PK Giridharan, P Ramesh Narayanan, and SVS Narayana Murty. 2014. "Microstructural investigations on ATIG and FBTIG welding of AA 2219 T87 aluminum alloy." *Applied Mechanics and Materials* 592:489–493.

Shyu, SW, HY Huang, KH Tseng, and CP Chou. 2008. "Study of the performance of stainless steel A-TIG welds." *Journal of Materials Engineering and Performance* 17 (2):193–201.

Sire, S, and S Marya. 2001. "New perspectives in TIG welding of aluminium through flux application." *Proceedings of the 7th International Symposium*, Kobe, Japan, 2001.

Sire, S, and S Marya. 2002. "On the selective silica application to improve welding performance of the tungsten arc process for a plain carbon steel and for aluminium." *Comptes Rendus Mecanique* 330 (2):83–89.

Sire, S, G Rückert, and S Marya. 2002. "Paper C-VII: Flux optimisation for enhanced weld penetration in aluminium contribution to FBTIG process." *WELDING IN THE WORLD-LONDON-* 46:207–218.

Takeuchi, Y, R Takagi, and T Shinoda. 1992. "Effect of bismuth on weld joint penetration in austenitic stainless steel." *Weld Journal* 71 (8).

Tanaka, M, T Shimizu, T Terasaki, M Ushio, and F Koshiishi. 2000. "Effects of activating flux on arc phenomena in gas tungsten arc welding." *Science and Technology of Welding and Joining* 5 (6):397–402.

Tathgir, S, A Bhattacharya, and TK Bera. 2015. "Influence of current and shielding gas in $TiO_2$ flux activated TIG welding on different graded steels." *Materials and Manufacturing Processes* 30 (9):1115–1123.

Tseng, KH. 2013. "Development and application of oxide-based flux powder for tungsten inert gas welding of austenitic stainless steels." *Powder Technology* 233:72–79.

Tseng, KH, and CY Hsu. 2011. "Performance of activated TIG process in austenitic stainless steel welds." *Journal of Materials Processing Technology* 211 (3):503–512.

Tseng, KH, and KL Chen. 2012. "Comparisons between $TiO_2$-and $SiO_2$-flux assisted TIG welding processes." *Journal of Nanoscience and Nanotechnology* 12 (8):6359–6367.

Tseng, KH, and NS Wang. 2014. "GTA welding assisted by mixed ionic compounds of stainless steel." *Powder Technology* 251:52–60.

Tseng, KH, and YJ Shiu. 2015. "Effect of thermal stability of powdered oxide on joint penetration and metallurgical feature of AISI 4130 steel TIG weldment." *Powder Technology* 286:31–38.

Vaishnani, S., J Sadhu, S Suthar, JJ Vora, and V Patel. 2017. "Development of Activated TIG welding Technology for Duplex Stainless steel 2205 for achieving sustainability." *Proceedings of 4th International Conference on Industrial Engineering* 2 (1): 458–462.

Vasantharaja, P, and M Vasudevan. 2012. "Studies on A-TIG welding of low activation ferritic/martensitic (LAFM) steel." *Journal of Nuclear Materials* 421 (1):117–123.

Vasantharaja, P, M Vasudevan, and P Palanichamy. 2015. "Effect of welding processes on the residual stress and distortion in type 316LN stainless steel weld joints." *Journal of Manufacturing Processes* 19:187–193.

Vora, JJ. 2017. "Development of flux assisted tungsten inert gas welding process for low activation ferritic martensitic steel." *Doctoral Thesis.*

Vora, JJ, VJ Badheka. 2016. "Experiences with A-TIG welding of RAFM steel: Comparative studies on the effect of carrier solvent." *Proceedings of Young Professional Seminar on Advances in Welding Technology and Automation*, Ahmedabad, India, 2016.

Vora, JJ, and VJ Badheka. 2015. "Experimental investigation on mechanism and weld morphology of activated TIG welded bead-on-plate weldments of reduced activation ferritic/martensitic steel using oxide fluxes." *Journal of Manufacturing Processes* 20:224–233.

Vora, JJ, and VJ Badheka. 2016a. "Experimental investigation on effects of carrier solvent and oxide fluxes in activated TIG welding of reduced activation ferritic/martensitic steel." *International Journal of Advances in Mechanical & Automobile Engineering* 3 (1):5. doi: 10.15242/IJAMAE.AE0316011.

Vora, JJ, and VJ Badheka. 2016b. "Improved penetration with the use of oxide fluxes in activated TIG welding of low activation ferritic/martensitic steel." *Transactions of the Indian Institute of Metals* 69 (9):1755–1764. doi: 10.1007/s12666-016-0835-6.

Vora, JJ, and VJ Badheka. 2017. "Experimental investigation on microstructure and mechanical properties of activated TIG welded reduced activation ferritic/martensitic steel joints." *Journal of Manufacturing Processes* 25:85–93.

Vora, JJ, V Patel, S Suthar, A Naik, D Kundal, R Patel, and S Das. 2017. "Investigation on the activated TIG welding of Cr-Mo-V steels." *Proceedings of 70th IIW International Conference (IIW 2017) at Shanghai, China*, 2017.

Xu, YL, ZB Dong, YH Wei, and CL Yang. 2007. "Marangoni convection and weld shape variation in A-TIG welding process." *Theoretical and Applied Fracture Mechanics* 48 (2):178–186.

Yang, C, S Lin, F Liu, L Wu, and Q Zhang. 2003. "Research on the mechanism of penetration increase by flux in A-TIG welding." *Journal Material Science Technology* 19 (1):225–227.

Yushchenko, KA, DV Kovalenko, and IV Kovalenko. 2006. "Peculiarities of A-TIG welding of stainless steel." *Trends in Welding Research: Proceedings of the 7th International Conference, May 16–20, 2005,* Callaway Gardens Resort, Pine Mountain, Georgia, USA.

Zhang, RH, JL Pan, and S Katayama. 2011. "The mechanism of penetration increase in A-TIG welding." *Frontiers of Materials Science* 5 (2):109.

Zhao, Y, G Yang, K Yan, and W Liu. 2011. "Effect on formation of 5083 aluminum alloy of activating flux in FBTIG welding." *Advanced Materials Research* 311:2385–2388.

# 12

## Friction Stir Welding of Aluminum Alloy: Principle, Processing, and Safety

**S. T. Azeez and E. T. Akinlabi**
*University of Johannesburg*

## CONTENTS

12.1   Introduction ........................................................................................262
12.2   Principle of Friction Stir Welding ..................................................262
12.3   FSW Processing ..................................................................................263
    12.3.1   Benefit of FSW ........................................................................264
    12.3.2   FSW of Aluminum Alloys .....................................................265
    12.3.3   Welding Tool Selection ..........................................................265
    12.3.4   Heat Generation Model .........................................................267
        12.3.4.1   Heat Generated from the Shoulder ........................267
        12.3.4.2   Heat Generated from the Probe .............................268
    12.3.5   Factors Affecting Weldability of Aluminum Alloy ..............269
12.4   Welding Hazards and Safety ............................................................270
    12.4.1   Accidents and Safety ..............................................................270
    12.4.2   Welding PPEs ..........................................................................270
    12.4.3   Safety Program for Welding Process .....................................270
    12.4.4   Clamping System .....................................................................271
        12.4.4.1   Advancing Side .......................................................271
        12.4.4.2   Retreating Side .......................................................272
    12.4.5   Hot Metal Safety Procedure ...................................................273
        12.4.5.1   Radiation ..................................................................273
        12.4.5.2   Electric shock ..........................................................274
        12.4.5.3   Hot Metal and Burns .............................................274
    12.4.6   Gases and Fumes .....................................................................275
    12.4.7   Welding Emission Preventive Steps ......................................275
12.5   Rotating Tool Safety ..........................................................................276
    12.5.1   Rotating Machine Safety .........................................................276
        12.5.1.1   Sources of Rotating Equipment Accident ..............276
    12.5.2   Machine Guarding ...................................................................277
12.6   Summary ..............................................................................................277
References .......................................................................................................277

## 12.1 Introduction

Welding activities are dangerous if safety program is not properly implemented. The operation can pose a physical threat to the operator if the operation procedure is not adhered to. There are more than four minor and major injuries per thousand reported all through service years as reported in OSHAcademy course 745 study guide, 2017. Also, there are several health risks connected to the welding process, for instance, infrared radiation and fumes. The risk is huge during fusion than in solid-state welding due to the melting of base metal and its alloy components. However, the choice of base materials and geometrical surface to be joined determines the level of threat. The friction stir welding (FSW) of similar light alloys/metals has been broadly studied, in relation to properties evolution (Mishra and Ma 2005). Furthermore, the FSW of light metal-matrix composites (MMCs) and dissimilar alloys are an important research area that has recently received wide attention (Debroy and Bhadeshia 2010; Murr 2010). Meanwhile, an improved FSW-welded joint has been produced between diverse alloys/metals, thick alloys, and MMCs (Chen 2009; Mofid et al. 2012; Sato et al. 2004; Sharifitabar and Nami 2001; Pan et al. 1996; Rodrigues et al. 2009). Until recently, there has been no scientific publication on safety and hazard prevention while operating the FSW machine. Understanding the working principle and processing parameters plays a crucial role in ensuring the safety of lives and sustainability of manufacturing technique.

## 12.2 Principle of Friction Stir Welding

FSW was invented in the early 1990s at The Welding Institute (TWI), United Kingdom (Thomas 1991). It is a reasonably new welding process, which has demonstrated its immense capability in joining metallic alloys that are conventionally known to be difficult to weld or are totally unweldable. Being a generic branch of a solid-state welding technique, FSW provides an enabling platform for welding alloys with comparatively flat geometry at an optimum productivity level (Azeez and Akinlabi 2018). Moreover, FSW is mainly suitable for joining light alloys/metals, especially aluminum-based MMC, similar and dissimilar alloys, respectively (Mishra and Ma 2005; Threadgill et al. 2009). Figure 12.1 shows the FSW machine during a welding operation.

However, FSW seems to be a reliable and distinctive welding technique, because it allows for the welding of dissimilar alloys, thereby avoiding the shortcomings of the fusion-welding process. The fabrication of dissimilar aluminum alloys is very attractive for most industrial applications, since we can join a less-expensive alloy with a more costly material when the need

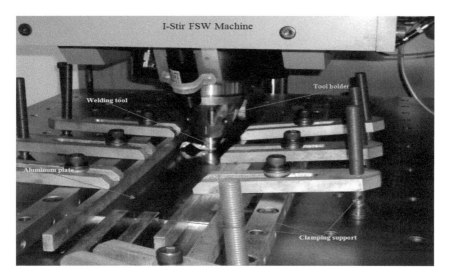

**FIGURE 12.1**
FSW welding machine.

arises. In fact, the joining of dissimilar materials might often be experienced in several engineering and structural-design instances, including aerospace, automotive, electronics, and shipbuilding, where the conventional fusion-welding technique is not appropriate, because of the huge disparities in chemical, physical, and mechanical properties between the materials/alloys to be fabricated (Debroy and Bhadeshia 2010; Guo et al. 2012; Murr 2010). Furthermore, challenges like porosity formation, slag inclusions, chemical reaction, and solidification cracking could occur during the traditional fusion welding of dissimilar alloys; however, good welds can be achieved in a few limited situations with a specific focus on weld-joints preparation, choice of filler metal, design, and process/welding parameters (Ellis 1996; Cam and Kocak 1998; Guo et al. 2012).

## 12.3 FSW Processing

FSW is a solid-state process of joining materials through severe plastic deformation. The process involves the plunging and traversing of a rotating cylindrical tool through the joint interface of the materials. The mechanical friction at the tool–material interface produces the heat required to soften the materials below its melting point. The integrity of the weld joint is dependent on the effectiveness of the selected process window (Azeez and Akinlabi 2018). The following terms are used to describe the FSW process:

*Tool rotation rate.* This is the rate of rotation of the welding tool. It makes a significant contribution to the total heat input and the material flow dynamics.

*Tool traverse speed.* This represents the speed at which the tool travels linearly along the welding joint, and it helps to determine the thermal cycle during welding.

*Tilt angle.* This is the angle between the spindle shaft and the plane normal to the workpiece. It normally has a numerical value between 0° and 3°.

*Plunge rate.* This is the rate at which the tool piece is plunged into the metallic plates. The heat generation and the plunge pressure are controlled during the initial tool insertion.

*Plunge depth.* This is the height at which the tool pin is inserted into the metallic plates from the top. The plunge depth is very important, while welding in the position controls the configurations.

*Plunge force.* This is the force exerted on the tool—when the tool shoulder is in contact with the surface of the metallic plates. This is very crucial when welding in a force-controlled configuration.

*Parent metal.* This is also referred to as the base metal. It is the section of the workpiece that is not affected by the heat input during the welding operation. The mechanical and metallurgical properties remain unchanged, but a thermal cycle is experienced.

*Heat affected zone (HAZ).* This is a region closer to the weld nugget, and a complete thermal cycle is experienced. A microstructural and mechanical evolution is noticed in this zone, although the workpiece is not plastically deformed.

*Thermomechanically affected zone (TMAZ).* This is a region of plastic deformation and microstructural changes. However, the heat input in this region induces plastic strain on the workpiece–without undergoing any recrystallization.

*Weld nugget.* This is a region of complete deformation and recrystallization that is also known as the stir zone. A severe plastic deformation is experienced by the materials; hence, there is a visible microstructural evolution (Azeez and Akinlabi 2018).

### 12.3.1  Benefit of FSW

FSW is referred to as the most important discovery in material fabrication in the past years. It is also referred to as a "green" technology, because of its environmentally friendly nature, power efficiency, and versatility. When compared with the conventional joining approach, it uses considerably less power—without using any consumables (i.e., flux or cover gas), and no toxic emission is released, making it an environmentally friendly welding method. In addition, FSW does not involve the melting of metal alloys; as a result, no filler metal is required; consequently, most of the aluminum grade can be fabricated or joined without any worry of solidification fracture, or composition compatibility, or any of the challenges associated with the conventional fusion welding. Furthermore, dissimilar aluminum grades and MMCs can be joined

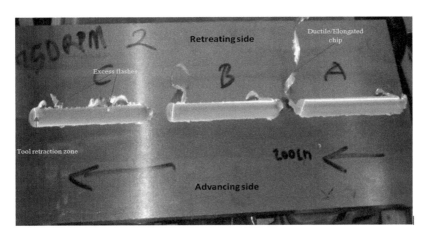

**FIGURE 12.2**
Friction stir welds formation.

with ease (Murr et al. 1998; Li et al. 1999, 2000). However, in traditional friction welding, limited to short axisymmetric components, the components can be pushed and rotated in opposite directions to one another to create a firm bond or joint (Cary 1979). FSW can be used to produce various geometrical shapes and joints, e.g., lap, butt, fillet, and T-butt shapes (Dawes and Thomas 1996).

### 12.3.2 FSW of Aluminum Alloys

The primary FSW parameters that can easily be controlled are the tool-movement parameters and the tool design. Nevertheless, other factors such as workpiece thickness, machine characteristics, and control mechanisms can equally affect the weld quality. The FSW parameters are also referred to as classic "path variables." This is because two similar FSW machines cannot perform in exactly the same way. Features like machine-tool eccentricity, control precision, and stiffness would be different for different machines. Nevertheless, in a repeated welding condition, any advancement in the process enhances the welding quality. The effect of processing parameters on weld formation is shown in Figure 12.2.

The tool rotation speed for the welding process is 750 rpm, while traverse speeds of 90, 120, and 150 mm/min, respectively, were used and alphabetically expressed as A, B, and C in Figure 12.2. The weld notation A is an elongated chip due to excess heat input while flashes in notation B and C are reduced. The three welds are good and are defect free.

### 12.3.3 Welding Tool Selection

The decision of FSW tool design parameters would be achieved from a synchronized microstructure and thermomechanical models. A collection of tool designs and tool behavior are developed, and the implication of these designs on the welding procedure explained. The tool materials that are usually used for FSW of aluminum alloys are tool steels, which have a mixture of high-temperature strength and toughness. At a high temperature, certain mechanical properties of the tool start to depreciate under service condition. Meanwhile, the aberration can be partially resolved using approximate temperature approach during FSW of aluminum alloys (~400°C–500°C) (Azeez and Akinlabi 2018). Figure 12.3 shows the FSW tool for aluminum alloy. In this manner, any tool steel with a temperature higher than 500°C is a perfect choice. The other viewpoint to consider is the maximum force accomplished during the plunging stage, which is a combination of tensile and compressive forces. These stresses can hasten tool breakage—particularly in a concave shoulder––that requires a tilted coordinate for its welding operation. A typical tool holder for aluminum and its alloy is shown in Figure 12.3. A definitive decision is, however, dictated by the following factors:

a. Tool run length;
b. Ease of material machining; and
c. Cost implication.

The other basic design factors are the dimension of the tool design (see Figure 12.4):

a. Tool shoulder diameter;
b. Tool pin diameter; and
c. Tool pin length.

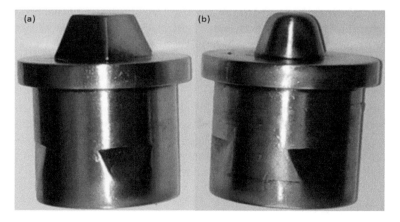

**FIGURE 12.3**
FSW Tool for aluminum alloys: (a) trapezoidal and (b) cylindrical.

**FIGURE 12.4**
FSW tool holder.

The choice of selecting a tool pin and tool shoulder diameter is, however, complicated by the following factors:

a. Workpiece geometry;
b. Joint reliability (i.e. joint strength measured relative to parent material strength); and
c. Required tool forces (i.e. mechanical behavior of tool).

The tool selection task is often less difficult if an optimum process window is considered in line with the materials properties. Hence, the life cycle of the tool prolonged and brittle failure prevented.

### 12.3.4 Heat Generation Model

In FSW, a conical tool shoulder described by a cone angle $\alpha$ is used in the heat generation model. A schematic representation of FSW tool design is shown in Figure 12.5.

### 12.3.4.1 Heat Generated from the Shoulder

The internal surface of modern FSW shoulder tool is mostly tilted. This is in contrast to the flat shoulder surface used in the early period of FSW invention. Tilted shoulder surface prevents material volume escape during plunging and enhances extrusion of material during the welding process (Schmidt et al. 2003). The heat produced from the tilted section of the shoulder tool is expressed as follows:

$$dQ = \omega r \, dF = \omega r^2 \tau_{\text{contact}} \, d\theta dr (1 + \tan \alpha) \qquad (12.1)$$

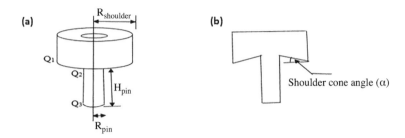

**FIGURE 12.5**
(a) Schematic illustration of heat generation location heat generation points and (b) schematic illustration of tool shoulder cone angle.

Integrating Equation 12.1 through the shoulder area across $R_{probe}$ to $R_{shoulder}$ gives the heat generated from the shoulder, $Q_1$

$$Q_1 = \int_0^{2\pi} \int_{R_{probe}}^{R_{shoulder}} \omega\tau_{contact} r^2 (1+\tan\alpha) \, dr \, d\theta \tag{12.2}$$

$$Q_1 = \frac{2}{3}\pi\tau_{contact}\omega\left(R_{shoulder}^3 - R_{probe}^3\right)(1+\tan\alpha) \tag{12.3}$$

### 12.3.4.2 Heat Generated from the Probe

The probe is a cylindrical shape part of the FSW tool. It comprises two major sections: the probe height and the probe radius. The heat produced from the probe is from two segments: $Q_2$ from the probe surface and $Q_3$ from the probe tip surface:

$$Q_2 = \int_0^{2\pi} \int_0^{H_{probe}} \omega\tau_{contact} R_{probe}^2 \, dz \, d\theta = 2\pi\tau_{contact}\omega R_{probe}^2 H_{probe} \tag{12.4}$$

Thus, we have

$$Q_3 = \int_0^{2\pi} \int_0^{R_{probe}} \omega\tau_{contact} r^2 \, dr \, d\theta \frac{2}{3}\pi\tau_{contact}\omega R_{probe}^3 \tag{12.5}$$

The total heat generated is the combination of the three individual contributions:

$$Q_{total} = Q_1 + Q_2 + Q_3 \tag{12.6}$$

where $Q_1$ is the heat generated below tool shoulder, $Q_2$ is the heat generated from tool-pin surface, and $Q_3$ is the heat generated from tool-pin tip:

$$Q_{total} = \frac{2}{3} \pi \tau_{contact} \omega \left( R_{shoulder}^3 - R_{probe}^3 + R_{probe}^3 + 3R_{probe}^2 H_{probe} \right) \qquad (12.7)$$

According to Khandkar et al. (2003), the total heat generation is expressed as follows:

$$Q_{total} = \frac{2}{3} \pi \tau_{contact} \omega \left( R_{shoulder}^3 + 3R_{probe}^2 H_{probe} \right) \qquad (12.8)$$

where

$Q_1$ = heat generation from the shoulder, W;
$Q_2$ = heat generation from the probe side, W;
$Q_3$ = heat generation from the probe tip, W;
$Q_{total}$ = total heat generation, W;
$R_{shoulder}$ = tool shoulder radius, m;
$R_{probe}$ = tool probe radius, m;
$H_{probe}$ = tool probe height, m;
$\mu$ = friction coefficient;
$\omega$ = tool angular rotation speed, rad/s;
$r$ = position along tool radius, m;
$\tau_{contact}$ = contact shear stress, Pa;
$\theta$ = angle, deg;
$\alpha$ = tool shoulder cone angle, deg;
$z$ = dimension along the rotation axis, m.

### 12.3.5 Factors Affecting Weldability of Aluminum Alloy

Weldability is the ability of materials to be joined together after being processed. The primary purpose of welding components into the desired shape is to improve the properties of the parent materials. The ease of processing and simplicity of procedure are key to weldability assessment. Thus, the capacity of a material to form a strong bond is the utmost target of most structural engineers. The stronger the materials affinity, the better the weld life cycle. Meanwhile, weldability depends on the following factors:

- Outline of the welding method.
- The chemical composition of parent materials.
- Alloying elements response to temperature changes.
- The degree of bulk expansion and contraction of the parent materials.
- Joint configuration or design layout.

## 12.4 Welding Hazards and Safety

A comprehensive understanding of welding hazards and methods of prevention is germane to ensuring the safety of lives and properties. Secondary hazards in welding workshop can be economically quantified, while severe accident like life-threatening injury or death is unquantifiable. In reducing welding hazards, the following steps can be taken:

- Ensuring that the workplace is properly ventilated.
- The equipment should be safe to operate.
- The operation should be done with reference to the manufacturer's manual and process guide.
- Ensure that appropriate personal protective equipment (PPEs) is selected for the operation.
- Evaluating the welding activities through a standard management system.

### 12.4.1 Accidents and Safety

Workshop accidents are avoidable mishap that can be managed if safety procedure can be properly defined. Also, the causes of hazards can either be due to human error or equipment condition. Manufacturers guide must strictly be studied and adhere to while operating an equipment. Training and retraining of welders are key to sustainable welding activities. In most engineering practice, timely reporting of abnormality as well as obtaining a work permit before carrying out welding activities are very important. Welding either in an enclosed space or at an altitude above 1.524 m must be accompanied by a technical support and under close supervision. Meanwhile, the first-aid kits must be accessible at every point in time (OSHAcademy 2017).

### 12.4.2 Welding PPEs

The primary purpose of PPE is for the protection of welding personnel and other people working in the workshop. Table 12.1 shows a summary of the protection part and the specific PPEs (OSHAcademy 2017).

### 12.4.3 Safety Program for Welding Process

An effective welding, cutting, and brazing program should protect workers from injury or death and property from fire, atmospheric contaminants, and other hazards during operations.

The main goal of safe welding program is to ensure the safety of lives, properties, and environment. It should also ensure the minimal release

**TABLE 12.1**

Summarized Personal Protective Equipment

| Protection | Personal Protective Equipment |
|---|---|
| Height or fall (i.e. above 1.524 m) | Safety belts, scaffolds, platforms, railings, and lifelines |
| Eye and face | Face and helmets, goggles, and spectacles |
| Hand | Gloves |
| Body | Protective clothing |
| Ear and nose | Earplug, nose cover, or respiratory aid |

of hazardous gases or no accidents occurs during the welding operation. If properly planned and implemented, injury, death, fire outbreak, air contaminations, etc. would be drastically eliminated. An effective implementation of a secured welding program involves a complete commitment from the top to the least member of the organization (OSHAcademy 2017). The following steps are important in getting the desired results:

- All the workers must be familiar with all the components of the program.
- The program must cut across all the departments, worksites, facilities, projects, etc. irrespective of the operation schedule or manpower.
- It should establish clear policies and objectives related to welding, cutting, brazing, and soldering safety. The program objectives and policies should be clearly stated in tandem with welding safety.

### 12.4.4 Clamping System

The clamping system is auxiliary accessories used for holding materials alloy and backing plate and to prevent the workpiece from swaying by welding pressure during processing. The backing plates play an important role during FSW of alloys, especially in the thermal cycle completion. The microstructural and mechanical properties evolution are directly connected to the physical and mechanical properties of backing plates. Meanwhile, there are two positions of material clamping during FSW operation, viz; (1) advancing side and (2) retreating side.

#### 12.4.4.1 Advancing Side

The direction of rotation of the tool pin surface and the direction of tool travel are in the same trajectory line. This is due to the tool's onward movement, and the material tends to flow backward, but the rotation of the pin surface resists that flow on the tool side.

### *12.4.4.2 Retreating Side*

The direction of rotation of the tool pin surface and that of the tool propagation are opposite in trajectory. The flow of the material on the tool pin surface is easier, and it supports the backward flow of the materials. Figure 12.6 shows a complete clamping system for aluminum plates.

The choice of selecting a backing plate is of significant importance, as it has a direct effect on the temperature gradient in the FSW process. Rosales et al. (2010) investigated the influence of backing-plate material on the FSW of butt welding of 3.2-mm-thick 2024-T351 and 4-mm-thick 6013-T6 and 3.2-mm-thick aluminum alloys. The three backing-plate materials copper, steel, and ceramic were investigated. A defect-free joint with a steel backing plate was observed, while excess flashes and tunnel defects were noticed with ceramic and copper-backing plates. Upadhyay and Reynolds (2012) investigated the influence of the thermal diffusivity of backing-plate materials and forging force on butt FSW of 6056-T451, 4.2-mm-thick aluminum alloy. A 3-mm-thick 2099 aluminum alloy, Al6XN, tool steel, ceramic materials, and Ti-6Al-4V were used as backing plates. They discovered that the apex, middle, and stir-zone temperatures for the ceramic-backing plate to be about 40°C, 35°C, and 25°C, and to be of higher magnitude than that for steel, Al6XN, and aluminum-backing plates, respectively. Zhang et al. (2013) experimented with the response of backing plates on butt FSW of 2024-T3 3.2-mm-thick aluminum alloy at a welding speed of 200 mm/min and a spindle speed of 600 rpm for granite, copper, and medium-carbon steel backing plates. The carbon steel backing plate was observed to give the highest welds efficiency of approximately 90%, while copper and granite-backing plates yielded significant lower weld efficiencies, respectively.

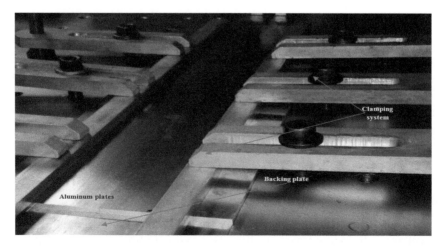

**FIGURE 12.6**
Clamping of aluminum alloys to the backing plate.

However, the existence of a tunnel defect in the welds was a factor for the reduction in weld strength for copper-backing plates, while HAZ softening could be the reason for the loss of weld strength for the granite-backing plate.

## 12.4.5 Hot Metal Safety Procedure

Hot work refers to all mechanical activities that involve heat generation either through chemical ignition or direct surface contact. The frictional heat generated during FSW makes the process to be classified as a hot working activity. The phase transformation and microstructural evolution of the welds depend on the amount of heat input into the parent materials. This is the basis of chemical, metallurgical, and mechanical properties enhancement. Recrystallization, precipitation, re-precipitation, change of state, and phase transition occur after the heat generated slightly rise above a specific threshold. This is the primary reason why safety practices are key to accident prevention as heat injuries are fatal and costly. The following practices are recommended for hot working activities:

- Hot work activities should not be performed close to combustible or flammable materials. This is to ensure a safe working environment that is nonhazardous.
- Functional and reliable fire extinguishers should be made available at every point in time.
- A smoke/fire detective wrist watch should readily available to sense smoke during hot work operation.

In addition, a secured welding procedure must strictly adhere to when operating FSW equipment. The following hazards should be prevented during FSW:

- Infrared radiations from the clamped plates.
- Rotating equipment.
- Sharp metal chips.
- Hot metal chips.
- Electric shock.

### 12.4.5.1 Radiation

The intensity of radiation emanating from FSW is quite minimal compared to arc welding. However, radiations from both processes need to be shielded from getting in contact with the welder. The head and eyes should be protected using a standard welding helmet and approved safety glasses.

- The hands, arms, face, and skin should be covered by radiation.
- Specified welding coat and gloves should be worn at all time.

### 12.4.5.2 Electric shock

In order to prevent the possibility of electrocution and shock in a fabrication workshop, electric cables and other electrical accessories/appliances are expected to be properly insulated. Safety ethics can now be established through PPE, e.g., safety gloves and boots. However, the cases of electric shock in FSW is minimal compared to the conventional welding technique. This is due to the total encapsulation of the FSW machine into a single-unit system. The mechatronics and the computational section are strategically embedded for a systematic welding operation. Welding either in a wet or enclosed location must be avoided.

### 12.4.5.3 Hot Metal and Burns

A severe skin burn can be experienced when a hot metal gets in contact with a welder. This is why is highly discouraged to handle hot metallic alloy with naked hands unless it has been air-cool or quenched in a liquid media. The appropriate safety measures for handling hot metals are by putting on a complete set of leather protection that has the following accessories:

- Ensure you put on sleeves or a jacket
- Welding apron
- Safety boots
- Welding spats
- Wear a gauntlet gloves
- Hat liners

Meanwhile, wounds or burn can be classified into three categories based on the sources of hazards:

- Thermal
- Radiation
- Chemical
- Electrical

Conversely, fusion welding through gas causes more skin hazard than the solid-state welding due to the high temperature involved and the penetration intensity of gas. The diffusion rate of most vapors or gases is extremely compared to that of liquids and solids. This is why exposure to radiation welding gases causes severe burns. It is encouraged to always put on high-flame-resistant clothes that will cover all parts of the body so as to shield the welder from the emitted infrared rays (OSHAcademy 2017).

### 12.4.6 Gases and Fumes

In solid-state welding, severe plastic deformation of metallic alloy takes place below the melting temperature. The solute component dissolution, precipitation, and recrystallization do not lead to the effervescent of alloying components. However, in fusion welding, the fluxes adopted welding processes generate hazardous vapors that cause irritation and damages to the throat, lungs, nose, and eyes (Azeez and Akinlabi 2016). The quantity of fumes and gases released into the environment is a function of the welding process window. In addition, to measure the degree of hazard due to exposure, a complete understanding of the welding principle and materials chemistry is required. The review of different welding processes, fume emission, and their effect on assessment indicator is stated in Table 12.2. Welding radiation and fumes contain toxic gases and particles from metal oxide that are produced during the welding process. The shielding gas composition, base metal, coating types, filler, exposure time, and welding procedure determine the quantity and composition of fumes emitted (Lyttle 2004).

### 12.4.7 Welding Emission Preventive Steps

In controlling welding fumes, various control measures are put in place for total exposure risk elimination or minimization. PPE and standard work practice are employed to reduce the degree of hazard to an infinitesimal extent. Below are few of the control measures that can be adopted in reducing

**TABLE 12.2**

A Review of Welding Emissions and Their Economic Importance

| Sources of Weld Emission | Emission Constituents | Economic Importance |
| --- | --- | --- |
| Arc welding | Carbon monoxide | Dizziness, muscular weakness, headache, loss in hearing, and overexposure can lead to death |
| Non-ferrous alloy and rod welding, wire coatings | Copper oxides | Metal fume fever, nose, eyes, and throat irritation |
| Emissions from rod coating | Hydrogen fluoride | Respiratory tract and bronchi irritation. Overexposure can lead to kidney, lung, and liver damage |
| Confined space welding | Oxygen deficiency | Unconsciousness, dizziness, and death |
| Welding magnesium-containing alloys | Magnesium oxide | Fume fever, nose, eyes, and throat irritation |
| Welding of non-ferrous alloy | Aluminum oxide | Severe pneumonitis |

exposure rate (Golbabaei and Monireh 2015 and OSHA Manual 1999). The following measures can be put in place in order to prevent exposure to emission:

- Global practices and work ethics
- Procedure modification
- Respiratory safeguard gadget

## 12.5 Rotating Tool Safety

The FSW machine is incorporated with tools that can either move transversely or revolve in a circle. The movement of the tools is powered by the mechanical impulse received from the electromechanical energy conversion unit of the equipment. The following terms are associated with FSW tool rotation:

*Tool rotation rate.* This is the rate of rotation of the tool. It makes a significant contribution to the total heat input and the flow of the materials.

*Tool pin.* This is also known as a probe. The tooling pin is inserted into the workpiece, and the horizontal material flow is affected when moving from front to back, along with the vertical material movement from top to bottom. The most prevalent types of pin profiles are conical or cylindrical. Tool pins can either be of spiral or threaded design. Furthermore, the material flow can be affected by placing the flutes. The entire probability of variations could be exceptionally large.

*Tool shoulder.* This is the region, where the tool is in contact with the surface of the workpiece. Tool shoulder scrolls enhance the material-flow process. Tool shoulders possess either positive or negative scrolls (Azeez and Akinlabi 2018).

### 12.5.1 Rotating Machine Safety

A rotating equipment if not properly guarded or encapsulated can cause severe injuries to the operating personnel. This often occurred whenever the power supplied to the rotating machine is uncontrollable. However, understanding the degree of risk and accidents associated with working on moving equipment is key toward ensuring the safety of lives and properties. A viable method of reducing workshop accidents is by introducing "guards" too sharp and rotating systems. In FSW, guards can be used to cover the revolving tool during plunging and traversing stages of the operation.

#### 12.5.1.1 Sources of Rotating Equipment Accident

The following are major causes of an accident while operating a moving equipment in the workshop:

- Untidy hair, jewelry, and loose clothe being entangled in moving parts.

- Materials removed from the machine when it is operational.
- Unintentional starting of the machine.
- Slipping and falling into an unguarded rotating tool nip.
- Direct contact with sharp edges, e.g., cutting blade.
- Adjusting the machine components while in operation.
- Operating a machine without proper authorization.
- Unavailability of the preventive maintenance system.

## 12.5.2 Machine Guarding

This simply means to encapsulate the dangerous part of a machine that can cause accident or injury. Meanwhile, guarding a machine also helps to eradicate hazards associated with operation and nip points, moving parts, or detached chips. In order to machine guard safety, then all the safety precautions and PPEs previously mentioned must be embraced (www.iitb.ac.in/safety/en/machine-safety). Nonetheless, the following safety precautions should be strictly adhered to when using a machine guard:

- Guards must be in good condition and position before the operation.
- Ensure cutting blades and tools are properly clean for easy operation.
- Do not use compressed air to clean the machine. This is to prevent dispersion of chips particles.

## 12.6 Summary

The safety of lives and sustainability of welding operation are crucial to the development of manufacturing industries. A coordinated welding program adopted can mitigate the severity of hazards encountered in a fabrication workshop. Thus, the economic, safety, and environment objectives of the process can be actualized with minimum resources and cost.

## References

Azeez, S.T. and Akinlabi, E.T. 2016. Toward improved health and quality of life: New goals for joining technology. *International Conference on Advances in Automotive Technologies*. Yildiz Technical University, Istanbul.

Azeez, S.T. and Akinlabi, E.T. 2018. Weld reliability and failure prediction of thermo-mechanical and metallurgical properties of friction stir welds of dissimilar aluminum alloys. PhD diss., University of Johannesburg.

Cam, G. and Kocak, M. 1998. Progress in joining of advanced materials. *International Materials Review* 43:1–44.

Cary, H.B. 1979. *Modern Welding Technology*. Upper Saddle River, NJ: Prentice-Hall.

Chen, T. 2009. Process parameters study on FSW joint of dissimilar metals for aluminum-steel. *Journal of Materials Science* 44:2573–2580.

Dawes, C.J. and Thomas, W.M. 1996. Friction stir process welds aluminum alloys: The process produces low-distortion, high-quality, low-cost welds on aluminum. *Welding Journal* 75:41–45.

Debroy, T. and Bhadeshia, H.K.D.H. 2010. Friction stir welding of dissimilar alloys–A perspective. *Science and Technology of Welding and Joining* 15:266–270.

Ellis, M.B.D. 1996. Joining of aluminum based metal matrix composites. *International Materials Reviews* 41:41–58.

Golbabaei, F. and Monireh, K. 2015. Air pollution in welding processes—Assessment and control methods. In F. Nejadkoorki (ed.). *Current Air Quality Issues*. London: IntechOpen.

Guo, J., Gougeon, P. and Chen, X.G. 2012. Microstructure evolution and mechanical properties of dissimilar friction stir welded joints between AA1100-B4C MMC and AA6063 alloy. *Materials Science and Engineering: A* 553:149–156.

Khandkar, M.Z.H., Khan, J.A. and Reynolds, A.P. 2003. Prediction of temperature distribution and thermal history during friction stir welding: Input torque based model. *Science and Technology of Welding and Joining* 8:165–174.

Li, Y., Murr, L.E. and McClure, J.C. 1999. Flow visualization and residual microstructures associated with the friction stir welding of 2024 aluminum to 6061 aluminum. *Materials Science and Engineering: A* 271:213–223.

Li, Y., Trillo, E.A. and Murr, L.E. 2000. Friction stir welding of aluminum alloy 2024 to silver. *Journal of Materials Science Letters* 19:1047–1051.

Lyttle, K. 2004. Optimizing consumable selection increases productivity, decreases fumes. *Gases & Welding Distribution* 48:45–47.

Machine Safety, www.iitb.ac.in/safety/en/machine-safety (accessed 18 December, 2017).

Mishra, R.S. and Ma, Z.Y. 2005. Friction stir welding and processing. *Materials Science and Engineering: R: Reports* 50:1–78.

Mofid, M.A., Abdollah-zadeh, A. and Ghaini, F.M. 2012. The effect of water cooling during dissimilar friction stir welding of Al alloy to Mg alloy. *Materials & Design* 36:161–167.

Murr, L.E. 2010. A review of FSW research on dissimilar metal and alloy systems. *Journal of Material Engineering and Performance* 19:1071–1089.

Murr, L.E., Li, Y., Flores, R.D., Trillo, E.A. and McClure, J.C. 1998. Intercalation vortices and related microstructural features in the friction-stir welding of dissimilar metals. *Materials Research Innovations* 2:150–163.

Occupational Safety and Health Administration (OSHA) 1999. *OSHA Technical Manual, Personal Sampling for Air Contaminants*. U.S. Department of Labor, Washington, DC.

OSHAcademy, Welding, Cutting, and Brazing Safety 2017. Course 745 study guide, www.oshatrain.org/courses/mods/745e.html (accessed 18 December, 2017).

Pan, C., Hu, L., Li, Z. and North, T.H. 1996. Microstructural features of dissimilar MMC/AISI 304 stainless steel friction joints. *Journal of Materials Science* 31:3667–3674.

Rosales, M.J.C., et al. 2010. The backing bar role in heat transfer on aluminium alloys friction stir welding. In L.G. Rosa and F. Margarido (eds.). *Materials Science Forum*. Zurich: Trans Tech Publ.

Rodrigues, D.M., Loureiro, A., Leitao, C., Leal R.M., Chaparro, B.M. and Vilaça, P. 2009. Influence of friction stir welding parameters on the microstructural and mechanical properties of AA 6016-T4 thin welds. *Materials and Design* 30:1913–1921.

Sato, Y.S., Yamashita, F., Sugiura, Y., Park, S.H.C. and Kokawa, H. 2004. FIB-assisted TEM study of an oxide array in the root of a friction stir welded aluminum alloy. *Scripta Materialia* 50:365–369.

Schmidt, H., Hattel, J. and Wert, J. 2003. An analytical model for the heat generation in friction stir welding. *Modelling and Simulation in Materials Science and Engineering* 12:143.

Sharifitabar, M. and Nami, H. 2001. Microstructures of dissimilar friction stir welded joints between 2024-T4 aluminum alloy and Al/Mg$_2$Si metal matrix cast composite. *Composite: Part B* 42:2004–2012.

Thomas, W.M. 1991. Friction stir butt welding. International Patent Application No. PCT/GB92/0220.

Threadgill, P.L., Leonard, A.J., Shercliff, H.R. and Withers, P.J. 2009. Friction stir welding of aluminum alloys. *International Materials Review* 54:49–93.

Upadhyay, P. and Reynolds, A.P. 2012. Effects of forge axis force and backing plate thermal diffusivity on FSW of AA6056. *Materials Science and Engineering: A* 558:94–402.

Zhang, Z., Li, W., Shen, J., Chao, Y.J., Li, J. and Ma, Y.E. 2013. Effect of backplate diffusivity on microstructure and mechanical properties of friction stir welded joints. *Materials & Design* 50:551–557.

# Index

## A

AC, *see* Alternating current (AC)
AC-TIG welding process, *see*
      Alternating current TIG
      (AC-TIG) welding process
Activated-tungsten inert gas (A-TIG)
      welding, 64, 66, 68, 238–240, 243
   aerodynamic drag forces, 254
   arc constriction mechanism, 248–249
   with different fluxes on steels,
      240–245
   electromagnetic forces, 253–254
   flux application methodology, 239
   forces in, 253
   oxide fluxes of, 252
   reversed Marangoni effect, 248–252
Adaptive process control, 93, 98
Additive manufacturing (AM), 77–78
   conventional manufacturing
      technologies, 78–79
   environmental benefits, 79
   resource efficiency, 78
   sector-wise implementation of, 78
Advanced high-strength steel
      substances (AHSS), 141
Aerodynamic drag forces, 253, 254
Aerospace industry, 42
A-FSW, *see* Assisted friction stir welding
      (A-FSW)
AHSS, *see* Advanced high-strength steel
      substances (AHSS)
AISI304 stainless steel, 66
Allied welding processes, 214
Al–Mg–Si–Cu alloys, 49
Alternating current (AC), 8, 9
Alternating current TIG (AC-TIG)
      welding process, 244
Aluminum alloys, 225–227
   clamping system, 271–273
   friction stir welding of, 265
   weldability of, 269
Aluminum–copper friction welding,
      226

AM, *see* Additive manufacturing (AM)
American Welding Society (2001), 102
Analysis of variance (ANOVA), 64, 65
ANN, *see* Artificial neural network
      (ANN)
ANOVA, *see* Analysis of variance
      (ANOVA)
Arc-based fusion welding processes, 213
Arc constriction mechanism, 248–249
Arc energy, 104–105
Arc stability, 16–18
Arc welding
   cable-type welding wire, 14–15
   gas metal, *see* Gas metal arc welding
      (GMAW)
   heat input in, *see* Heat input, in arc
      welding
   metal transfer modes
      gas metal, 25–29, 127
      twin-wire indirect, 14
   submerged, *see* Submerged arc
      welding (SAW)
   twin-wire indirect, 13–14
Artificial neural network (ANN), 67, 68,
      70
Assisted friction stir welding (A-FSW),
      37–38
   electrically assisted FSW, 39–41
   gas torch—FSW, 38
   gas tungsten arc welding—FSW, 39
   induction heating tool-assisted FSW,
      40, 41
   laser—FSW, 39, 40
   plasma—FSW, 38, 39
   ultrasonic energy-assisted FSW,
      41–42
A-TIG welding, *see* Activated-tungsten
      inert gas (A-TIG) welding
Automated cryogenic calorimeter, 107,
      111, 124
Automative industry, 81
Automobile industries, 42
AZ31, *see* Magnesium alloy (AZ31)

**B**

Base metals (BMs), 160
BBD, *see* Box–Behnken design (BBD)
Bead geometry, 8, 30, 66, 127, 128, 129, 133
BMs, *see* Base metals (BMs)
Bobbin tool friction stir welding (BTFSW), 45
Box–Behnken design (BBD), 66, 70
BTFSW, *see* Bobbin tool friction stir welding (BTFSW)

**C**

Cable-type welding wire (CWW) arc welding, 14–15
Calorimeter/calorimetry
  cryogenic, 107, 111–112
  intrinsic errors on, 106–110
  liquid nitrogen, 123
Casting weld repairing, 49
Central composite design (CCD), 66, 70
Ceramics, friction welding of, 229–231
Clamping system, of aluminum alloys, 271–273
CMT, *see* Cold metal transfer (CMT)
Coefficient of thermal expansion (CTE), 230
Cold metal transfer (CMT), 84, 92
Composite manufacturing, surface and bulk, 48–49
Contact tube-to-work piece distance (CTWD), 4
Cryogenic calorimeter, 107, 111–112, 115
CTE, *see* Coefficient of thermal expansion (CTE)
CTWD, *see* Contact tube-to-work piece distance (CTWD)
CWW arc welding, *see* Cable-type welding wire (CWW) arc welding

**D**

DC, *see* Direct current (DC)
Deep penetration, 2
DE-GMAW, *see* Double-electrode GMAW (DE-GMAW)

Direct current (DC), 8, 9
Direct drive friction welding, 216
Direct metal deposition (DMD), 82
Dissimilar materials joint design, 212–214
Double-electrode GMAW (DE-GMAW), 12–13
Double-side spot welding, 177–178

**E**

EAFSW, *see* Electrically assisted friction stir welding (EAFSW)
EBAM, *see* Electron beam additive manufacturing (EBAM)
EBM, *see* Electron beam melting (EBM)
Elapsed time and trajectory, intrinsic error due to, 114–117
Electrically assisted friction stir welding (EAFSW), 39–41
Electric shock, friction stir welding, 274
Electromagnetic characteristics of arcs, 14
Electromagnetic forces, 253–254
Electromagnetic interaction (EMI), 8, 9
Electron beam additive manufacturing (EBAM), 79, 80
Electron beam freeform fabrication (EBF3), 83, 84
Electron beam melting (EBM), 81
EMI, *see* Electromagnetic interaction (EMI)

**F**

FB-TIG welding, *see* Flux-bounded TIG (FB-TIG) welding
FEM, *see* Finite-element method (FEM)
Ferrous alloys, 181–183
Fillet weld, 2, 44–45
Finite-element method (FEM), 87
Flat roller, 96, 97
Flux application methodology, 239
Flux-bounded TIG (FB-TIG) welding, 245–246
Flux-zoned TIG (FZ-TIG), 246–248
FPDZ, *see* Fully deformed zone (FPDZ)

Frcition welding (FW)
  aluminum alloys, 225–227
  of ceramic materials, 229–231
  copper and its alloys, 229
  definition of, 214
  of dissimilar materials combination, 222–223
    burn-off and reaction layer thickness, 223–224
    deformation behavior, 223
    mixing at interface, 224
  forging stage, 219–220
  friction stage, 218–219
  heat-affected zones during, 221–222
  heat generation in, 221
  heating time, 221
  orbital, 218
  of polymers, 231
  procedural steps, 214–215
  rotary, 216–218
  speed and pressure, 220
  steels, 227–229
Friction spot welding (FSpW), 176–177
Friction stir channeling (FSC), 46
Friction stir processing (FSP), 46–47
  casting and fusion weld repairing, 49
  grain refinement, 47–48
  superplasticity, 48
  surface/bulk composite materials, 48–49
Friction stir spot welding (FSSW), 42–44, 161
  advantages of, 160
  ferrous alloys, 181–183
  lap-shear
    stack and temperature, 171
    tensile properties, 172
    tool tilt angle and dwell time on, 172, 173
  microhardness profiles, 170
  microstructural evolution, 161–162
    cracking, 165–167
    hook formation, 163–164
    intermetallic formation, 166–168
    local melting, 165, 166
    material flow, 162–163
  nonmetals, 183–185
  pin/probe and shoulder, 161
  process variants

double-side spot welding, 177–178
    external heating, 178, 180
    particle addition, 180
    pin-less tool, 178, 179
    refilling spot welding, 176–177
  rotational speed, 169, 170
  stir zones, 169, 176
  tool design effect
    geometries, 172, 174–175
    macroscopic cross section, 172, 174
Friction stir welding (FSW), 160, 262
  advantages and limitations, 35–37
  of aluminum alloys, 265
    factors affecting weldability, 269
  benefits of, 264–265
  electrical-current-aided, 41
  heat generation model
    from probe, 268–269
    shoulder tool, 267–268
  of high-strength steels, 139
    advantages, 141
    defects in, 154–155
    equipment requirements for, 148–149
    hydrocarbon-based fuel sector, 141–142
    microhardness evolution, 149–150
    microstructural evolution, 151–154
    microstructural zones in, 143–144
    oxide scattering reinforced ferrite steels, 142
    process parameters in, 150–152
    refractory metal tools, 144–146
    reinforcing constructional materials, 143
    schematic setup of, 140
    super-abrasive tools, 146–147
    tool wear, 147–148
  plasma-assisted, 39
  principle of, 262–263
  processing steps, 263–264
  rotating tool safety, 276–277
  tool selection, 265–267
  variants of, 37
  welding hazards and methods
    accidents and safety, 270
    clamping system, 271–273
    emission preventive steps, 275–276

Friction stir welding (FSW) (*cont.*)
  gases and fumes, 275
  hot metal safety procedure,
    273–275
  personal protective equipment,
    270, 271
  safety program for, 270–271
  welding machine, 263
  working principle of, 34–35
FSC, *see* Friction stir channeling (FSC)
FSpW, *see* Friction spot welding (FSpW)
FSSW, *see* Friction stir spot welding
    (FSSW)
FSW, *see* Friction stir welding (FSW)
Fully deformed zone (FPDZ), 230
Fusion welding, 1
FW, *see* Frcition welding (FW)
FZ-TIG, *see* Flux-zoned TIG (FZ-TIG)

**G**

GA, *see* Genetic algorithm (GA)
Gas metal arc-welding (GMAW), 1, 3,
    126, 127
  additive manufacturing, 83, 84
  double-electrode, 12–13, 92
  regulated metal deposition
    technique
    metal-transfer modes, 23–25
    modified short-circuit, 25–28
    root pass, 27–31
  tandem, 8–9
  twin-wire system for, 6–7
Gas torch friction stir welding (GFSW),
    38
Gas tungsten arc welding (GTAW), 83,
    129, 133
Gas tungsten arc welding friction stir
    welding (GTAFSW), 39
Genetic algorithm (GA), 67–69
GFSW, *see* Gas torch friction stir
    welding (GFSW)
GMAW, *see* Gas metal arc-welding
    (GMAW)
GRA, *see* Grey relational analysis (GRA)
Grain refinement, 47–48
Green technology, 264
Grey relational analysis (GRA), 65
Grey relational grade (GRG), 65

GRG, *see* Grey relational grade (GRG)
Grooved roller, 96
Gross heat input, intrinsic errors due to,
    110–112
  elapsed time and trajectory, 114–117
  test plate dimensions
    length, 122–124
    thickness, 117–119
    width, 119–123
  weld bead length, 112–114
GTAFSW, *see* Gas tungsten arc
    welding friction stir welding
    (GTAFSW)
GTAW, *see* Gas tungsten arc welding
    (GTAW)

**H**

Heat-affected zones (HAZs), 35, 64,
    67–68, 143, 152–153, 194, 196,
    198, 206
  definition of, 162, 264
  during friction welding, 221–222, 224
  nonuniform, 217
  width of, 228
Heat flux, inside plate during welding,
    104–105
Heat input, in arc welding, 101–102
  calculate welding energy, methods
    to, 125
  gross, *see* Gross heat input
  heat flux inside plate, 104–105
  intrinsic errors, 103, 106–110
  mean/RMS values, 126
    absorbed heat for, 134
    bead geometry modeling, 127–129,
      131–132, 133
    calorimetric methods, limitations
      of, 133
    pulse-based current, 129
Higher dissolving temperature
    materials (HTMs), 140
  friction stir welding in
    refractory metal tools, 144–146
    super-abrasive tools, 146–147
Higher level of nitrogen (HNS), 142
HTMs, *see* Higher dissolving
    temperature materials (HTMs)
Hydrocarbon-based fuel sector, 141–142

**I**

i-FSW, *see* Induction heating tool-
    assisted friction stir welding
    (i-FSW)
IMCs, *see* Intermetallic compounds
    (IMCs)
Induction coil, 40–41
Induction heating tool-assisted friction
    stir welding (i-FSW), 40–41
Inertia friction welding process, 217
Inertial drive welding, 216
In-situ bead profile sensing, 93
In-situ rolling controls, 94–97
Intermetallic compounds (IMCs),
    166–168, 220, 227
Intermetallic layer (IML), 226
Inverted profiled roller, 96
ISO/TR 18491:2015 standard, 125

**K**

Kissing bonds cracks, 154

**L**

LAM, *see* Laser additive manufacturing
    (LAM)
Lap-shear (LS)
    stack and temperature, 171
    tensile properties, 172
    tool tilt angle and dwell time on, 172,
        173
Laser additive manufacturing (LAM),
    79, 80
Laser-based direct metal deposition
    (LBDMD), 91–92
Laser beam welding parameters, 65, 66
Laser engineering net shaping (LENS),
    82
Laser friction stir welding (LFSW), 39,
    40
Laser powder deposition (LPD), 82
LASERTEC 65, 94
LBDMD, *see* Laser-based direct metal
    deposition (LBDMD)
LENS, *see* Laser engineering net shaping
    (LENS)
LFSW, *see* Laser friction stir welding
    (LFSW)

LFW, *see* Linear friction welding
    (LFW)
Linear friction welding (LFW)
    advantages and disadvantages of,
        192–193
    block diagram of, 192
    definition of, 191
    phases of, 191–192
    titanium alloys
        fatigue properties and torsional
            strength, 205–206
        hardness profiles, 201–204
        impact and fracture toughness,
            205–206
        macroscopic features, 193–194
        residual stresses, 199–202
        tensile properties, 205
        texture formation, 197–199
        Ti64 joint, 194–197
Liquid nitrogen calorimeter, 123
Liquid penetration induced (LPI), 165
Loss function, 58
LPD, *see* Laser powder deposition (LPD)
LS, *see* Lap-shear (LS)
LUMEX Avance-25/Avance-60, 94

**M**

Machine guarding, 277
Magnesium alloy (AZ31), 48, 49
Magnetic fields, behavior of, 8, 9
MAT, *see* Medial axis transformation
    (MAT)
Material efficiency, of AM process, 78
MATLAB® software, 68
Mazda Motor Corporation, 42
Mechanical fastening, 214
Mechanically assisted droplet transfer,
    24
Medial axis transformation (MAT),
    85, 86
Melting point depression, 246
Melting rate (MR), 126
Meta-heuristics optimization method,
    66–69, 71
Metal-cored wires, 31
Metal-matrix composites (MMCs), 180,
    262
Metal–metal friction joint, 231

Metal transfer modes, arc welding
   gas metal, 25–29, 127
   twin-wire indirect, 14
Microstructural evolution, friction stir
     spot welding, 161–162
   cracking, 165–167
   hook formation, 163–164
   intermetallic formation, 166–168
   local melting, 165, 166
   material flow, 162–163
MIG welding, twin-wire, 10
Miller Electric Company, 25
Milling-assisted wire and arc additive
     manufacturing, 94, 95
MMCs, *see* Metal-matrix composites
     (MMCs)
Multiple-wire welding, 2–4, 18
   cable-type welding wire arc welding,
     14–15
   gas metal arc welding
     double-electrode, 12–13
     tandem, 8–9
   research trends in, 16–17
   submerged arc-welding
     double-electrode, 12–13
     metal powder-assisted, 15–16
   transferred ionized molten energy-
     twin, 11–12
   twin-wire indirect arc welding
     novel technology in, 13–14
   twin-wire welding, 2, 4–8, 18
     power source synchronization,
     10–11
Multi-walled carbon nanotubes
     (MWCNTs), 48–49

**N**

National Pingtung University of
     Science and Technology
     (NPUST), 243
Non-arc-based fusion welding
     processes, 213
Nonconsumable rotating tool, 34–35
Nonmetals, friction stir spot welding,
     183–185
NPUST, *see* National Pingtung
     University of Science and
     Technology (NPUST)

**O**

ODS ferrite steels, *see* Oxide
     scattering reinforced (ODS)
     ferrite steels
Optimization of welding parameters,
     55–56
   meta-heuristics methods, 66–69
   process and methods, 57
   response surface methodology, 66
   Taguchi optimization, 57–66
Orbital friction welding, 218
Orthogonal array (OA), *see* Taguchi
     optimization method
Overhanging features, of welding-
     based additive manufacturing,
     86–92, 97
Oxide scattering reinforced (ODS)
     ferrite steels, 142

**P**

Parent material (PM), 195, 205
Parent metal, 264
Pareto fronts, 69, 71
Part flexibility, of AM process, 79
Path planning, of additive
     manufacturing, 84, 85, 97
PAW, *see* Plasma arc welding (PAW)
PAZ, *see* Plastically affected zone (PAZ)
PBF, *see* Powder bed fusion (PBF)
PCBN tools, *see* Polycrystalline cubic
     boron nitride (PCBN) tools
PCD, *see* Polycrystalline diamond (PCD)
Personal protective equipment (PPE),
     270, 271
PFSW, *see* Plasma friction stir welding
     (PFSW)
Pinch rollers, 96, 97
Pin-less tool, 178, 179
Plasma arc welding (PAW), 128, 249
Plasma friction stir welding (PFSW),
     38, 39
Plastically affected zone (PAZ), 196
Polycrystalline cubic boron nitride
     (PCBN) tools, 146–147, 149
Polycrystalline diamond (PCD), 146
Polymers, 79
   friction welding of, 231

Post-weld heat treatment (PWHT), 196, 198, 201, 203, 206
Powder bed, additive manufacturing, 79–83, 87, 94
Powder bed fusion (PBF), 80, 81, 87
Power sources synchronization, twin-wire welding, 10–11
Profiled roller, 96, 97
Pulsed current, 10, 12, 129
PWHT, *see* Post-weld heat treatment (PWHT)

**R**

Radiation, friction stir welding, 273–274
Reduced activation ferritic martensitic steels (RAFM), 245
Refilling spot welding, 176–177
Refractory metal tools, 144–146
Regulated metal deposition (RMD)
    applications of, 28–31
    cycle, 25–26
    gas metal arc-welding
        metal-transfer modes, 23–25
        modified short-circuit, 25–28
        root pass, 27–31
Resistance spot welding (RSW), 159–160
Response surface methodology (RSM), 66, 70
Reversed Marangoni effect, 248–252
RMD, *see* Regulated metal deposition (RMD)
Rolling-assisted welding additive manufacturing, 94–97
Root pass welding, 27–31
Rosenthal model, 122, 123
Rotary friction welding, 214–218
RSM, *see* Response surface methodology (RSM)
RSW, *see* Resistance spot welding (RSW)

**S**

SAW, *see* Submerged arc-welding (SAW)
Seebeck calorimeter, 107
Selective laser melting (SLM), 81, 94
Selective laser sintering (SLS), 81, 82
Self-reacting friction stir welding (SRFSW), 45

Short-circuiting mode, 25
Single-wire welding, 2
SLM, *see* Selective laser melting (SLM)
SLS, *see* Selective laser sintering (SLS)
Solid-state processes, 214
Solid-state welding, 275
Spiral tool-path, 85
SRFSW, *see* Self-reacting friction stir welding (SRFSW)
SSFSW, *see* Stationary shoulder friction stir welding (SSFSW)
Stationary shoulder friction stir welding (SSFSW), 44–45
Stir zones (SZs), 35, 36, 150, 169
Submerged arc-welding (SAW), 1
    double-electrode, 12–13
    metal powder-assisted, 15–16
    twin-wire system for, 6–8
Super-abrasive tools, 146–147
Superplasticity, 48
SZs, *see* Stir zones (SZs)

**T**

Taguchi optimization method, 57–67, 70
Tandem welding system, 2, 8–9
Test plate dimensions, intrinsic errors due to
    length, 122–124
    thickness, 117–119
    width, 119–123
Thermal-related welding models, 110
Thermo-mechanically affected zone (TMAZ), 35, 143, 153, 162, 166, 195, 196, 202, 204, 205, 264
Thermomechanical processing (TMP), 48
TIG welding process, *see* Tungsten inert gas (TIG) welding process
TIME-twin, *see* Transferred ionized molten energy (TIME)-twin
Ti-6Al-4V (Ti64), 191, 193
Ti64 linear friction weld, 194–197
Titanium alloys, linear friction welding
    fatigue properties and torsional strength, 205–206
    hardness profiles, 201–204

Titanium alloys, linear friction welding (*cont.*)
impact and fracture toughness, 205–206
macroscopic features, 193–194
residual stresses, 199–202
tensile properties, 205
texture formation, 197–199
Ti64 joint, 194–197
TMAZ, *see* Thermo-mechanically affected zone (TMAZ)
TMP, *see* Thermomechanical processing (TMP)
Trail wires, 1, 10
Transferred ionized molten energy (TIME)-twin, 11–12
Trial-and-error method, 64
Triple-wire system, 7–8
Tungsten arc welding, 39
Tungsten inert gas (TIG) welding process, 29, 237–238
activated TIG
arc constriction mechanism, 248–249
with different fluxes on steels, 240–245
flux application methodology, 239
depth-enhancing mechanisms, 247–248
flux-bounded TIG, 245–246
flux-zoned TIG, 246–247
Tungsten–rhenium tools, 145–146
TWI, *see* The Welding Institute (TWI)
TWIA welding, *see* Twin-wire indirect arc (TWIA) welding
Twinning-induced plasticity steel (TWIP), 181, 182
Twin-wire indirect arc (TWIA) welding, 13–14
Twin-wire welding, 2, 4–8, 18
power source synchronization, 10–11
TWIP, *see* Twinning-induced plasticity steel (TWIP)

U

Uber Stir Technologies, 146
Ultrasonic energy-assisted friction stir welding (UAFSW), 41–42

Ultrasonic transducer, 41, 42

W

WAAM, *see* Wire and arc additive manufacturing (WAAM)
Water calorimeter, 106
Water-cooled stationary anode calorimeter, 108–109
Weighted additive utility function (WAUF), 67–69, 71
Weldability, of aluminum alloy, 269
Weld bead geometry, *see* Bead geometry
Weld bead length, intrinsic error due to, 112–114
Weld center zone (WCZ), 195–196, 198, 201–204, 206
Welding-based additive manufacturing (WAM)
considerations
overhanging features, 86–92
path planning, 84–86
developments
advance power sources, 92–93
hybrid processes, 94
in-situ sensing and adaptive process control, 93
rolling-assisted, 94–97
techniques
powder-based AM technology, 79–83
wire-feed-based, 83–84
The Welding Institute (TWI), 139, 160, 262
Welding procedure specifications (WPSs), 102
Weld nugget (WN), 35, 67–68, 143, 151
Weld root flaw cracks, 154
Wire and arc additive manufacturing (WAAM), 79, 80, 83, 86
adaptive control of, 93
milling-assisted, 94, 95
rolling-assisted, 94–97
Wire and laser additive manufacturing (WLAM), 83
Wire-feed-based welding additive manufacturing, 83–84
WN, *see* Weld nugget (WN)
Wormhole, 46

WPSs, *see* Welding procedure
        specifications (WPSs)

**X**

X-ray quality weld, 29, 30

**Z**

Zigzag tool-path generation, 85